Ethical Reasoning for a Data-Centered World

By

Rochelle E. Tractenberg, PhD, MPH, PhD, PStat®, FASA, FAAAS

Ethical Reasoning for a Data-Centered World

By Rochelle E. Tractenberg, PhD, MPH, PhD, PStat®, FASA, FAAAS

This book first published 2022

Ethics International Press Ltd, UK

British Library Cataloguing in Publication Data

A catalogue record for this book is available from the British Library

Print Book ISBN: 978-1-80441-078-3

eBook ISBN: 978-1-80441-079-0

Dedication

This book is dedicated to my fellow practitioners, and to instructors who have an interest in integrating ethical reasoning, and ethics, into their courses on statistics, computing, and data science. I know that students and practitioners who work with data directly or indirectly may share this interest at different times in their careers and training; I hope the book serves instructors, learners, and practitioners equally well in their commitment to ethical engagement with data.

Acknowledgements

This is my first book, the first draft of which was completed during a sabbatical from Georgetown University (2019). My ethical reasoning mentor, Father Kevin FitzGerald SJ, PhD, PhD, was instrumental in forging my commitment to ethical reasoning for teaching, learning, and enabling others to develop an ethical professional identity. Father Kevin's friendship, optimism, and encouragement have shone some of the brightest lights on my two decades (so far!) at Georgetown.

While the book is focused on ethical reasoning, and can be used to build its requisite knowledge, skills, and abilities generally, it is structured around statistics, computing, and data science. While all errors are definitely my own, the work benefitted greatly from frequent, fun, and challenging discussions with Donna LaLonde of the American Statistical Association about the obstacles that instructors in statistics and data science might face when they seek to integrate ethical reasoning and ethical content into their courses. (The conversations about dogs and running were also very helpful!) Donna's relentlessly concrete suggestions and honest feedback have made her a great friend as well as an awesome sounding board. I am lucky to have the benefit of her experience, expertise, and encouragement.

The Committee on Professional Ethics (COPE) of the American Statistical Association, which I had the privilege of vice-chairing (2014-2016) and chairing (2017-2019), was formative in my commitment to writing this book in order to help all instructors to utilize the ASA Ethical Guidelines for Statistical Practice. I am grateful to Howard Hogan, fellow Camarillo resident, for his faith in me and encouragement to fully engage with the Committee's charter.

Stewardship of the profession and discipline of science has been a feature of my professional identity for decades. I owe a debt of gratitude to the Carnegie

Initiative on the Doctorate and Chris Golde in particular for bringing the construct of disciplinary stewardship into being. Dr. Golde, and Dr. Chris Rios, facilitated my thinking about expanding stewardship beyond doctoral training and to the profession. I hope this book can serve to engage readers with disciplinary and professional stewardship as they learn to reason ethically in the modern, data-centered, world.

Table of Contents

List of Cases

Abbreviations:

DEFW is the Data Ethics Framework

DSEC is the Data Science Ethics Checklist

ASA is the American Statistical Association

ACM is the Association of Computing Machinery

Page	Case Title	Relevant Principles
321	**Case:** Results suggest that some people who are the source of data your organization wants to use (e.g., Facebook) are more susceptible to messaging (e.g., advertisements and fake news items) than others. You interpret this as signalling a need for caution/care in what your system does next with this data, and you include this in all your communication – in order to limit bias, ensure that no stakeholders are misled, and support valid conclusions resulting from your statistical, computing, and data science practice. Instead, you find reports of your work interpret the results without any caveats, and remove any suggestions that caution or sensitivity analyses may be needed.	ASA: A, B, C, D,E, F, G, H **ACM:** 2
331	**Case:** You are told **not** to document your work. When you do (because that's what the ethical practitioner does) your boss/supervisor returns it to you with the direction, "fix this".	ASA: A, B, C, E, F, G, H **ACM:** 1, 2, 3, 4
340	**Case:** You submit your complete and correct report of your scraping algorithm – including identification of the removal of your built in, opt-in consent to contribute data; the lack of consent accompanying data to be analyzed; and the lack of your recommendations in interpretations for limiting bias. You later discover that none of this documentation was included in the final report, but the final report is shared with stakeholders as if it is complete and correct.	ASA: A, B, C, D, E, F, G, H **ACM:** 1, 2, 3, 4
355	**Case:** Leadership informs your team that they bought an algorithm that you will be using to scrape data. But first, they want you to take off all the consent pop-ups, because "that ruins the user experience" and "adds personal data we will only need to strip off to preserve confidentiality".	ASA: A, B, C, E, G, H **ACM:** 1, 2, 3, 4

Introduction

The subject of the book is *ethical reasoning*: how to do it, why learn to do it, and how to teach and learn to do it, and document that it has been learned (and improved). Ethical reasoning is its own set of knowledge, skills, and abilities (KSAs (Santa Clara University (no date); Tractenberg & FitzGerald, 2012; Tractenberg et al. 2017)). These KSAs are learnable and improvable, and can be deployed to ensure ethical practice (when there is no/before there is an ethical problem about which a decision has to be made) as well as when a decision about what to do (ethically) is required. Thus, learning to reason ethically – rather than "learning the Ethical Guidelines and/or Code of Ethics" – will promote "…the skill, good judgment, and polite behavior that is expected from a person who is trained to do a job well" more generally, and more universally (see Rios et al. 2019).

The academic level at which the book is targeted is anyone who is preparing to engage in statistics or data science in the course of their work – whether that is their main task or simply a toolset they will utilize. This book introduces and discusses ethical reasoning as a learnable, improvable skill set that an individual can learn themselves, and document, *and* which an instructor can teach and assess (and develop themselves). The context of the book is "the data-centered world", wherein "quantitative practice" is any person's applications of statistics, data science, or some combination of these. Importantly, an individual who does one "simple analysis" is engaging in quantitative practice, although not to the extent that someone with a job title "statistician" does; as you will see in later chapters, practice standards are relevant for both of these individuals – making this book, and its application of those practice standards across tasks, relevant right across quantitative practice – however extensive or limited it may be.

The material and examples should be accessible to advanced undergraduate or graduate students, as well as practitioners. This book presents the current (2022) ASA Ethical Guidelines for Statistical Practice[1], which are emphatically relevant for <u>any</u> person analyzing data, whether they are a researcher/scientist or not; and whether or not the practitioner self-identifies as a statistician. Also presented are the current (2018, draft 3) Code of Ethics of the Association of

---

[1] https://www.amstat.org/ASA/Your-Career/Ethical-Guidelines-for-Statistical-Practice.aspx

Computing Machinery[2]. The book could be used for a stand-alone course, as an adjunct text for a consulting course, or for use alongside an internship and/or capstone course (for advanced undergraduates or graduate students) in any discipline or program where data are utilized.

How Ethical Reasoning Figures in Professional Identity Development

Figures 1A and 1B[3] show two alternative models of professionalism specific to quantitative sciences. The figures are adapted by the author from a model for professionalism in the health professions (Figure B, Arnold & Stern, 2006). In the figure below, the five key elements of competent work by statisticians (identified by Rosnow & Rosenthal, 2011) are shown as pillars supporting "professionalism": transparency, precision, accuracy, accountability, and logical and scientific justification <for work>. There is no mention of ethical knowledge or reasoning in any discussion of professionalism among those who use statistics and data science –whether this is a primary or secondary/ancillary part of their work. The American Statistical Association (ASA) Ethical Guidelines for Statistical Practice (or GLs) and the Association for Computing Machinery (ACM) Code of Ethics (or CE) implicitly blend these two constructs – ethics and professionalism. Both of these professional practice standards are also explicit in their applicability to any person, irrespective of job title or educational background, that will utilize the tools, methods, and constructs of these professions.

---

[2] https://www.acm.org/about-acm/code-of-ethics

[3] The figures and descriptive paragraphs are excerpted from Tractenberg RE. (2013). Ethical Reasoning for Quantitative Scientists: A Mastery Rubric for Developmental Trajectories, Professional Identity, and Portfolios that Document Both. Proceedings of the 2013 Joint Statistical Meetings, Montreal, Quebec, Canada. Pp. 3959-3973.

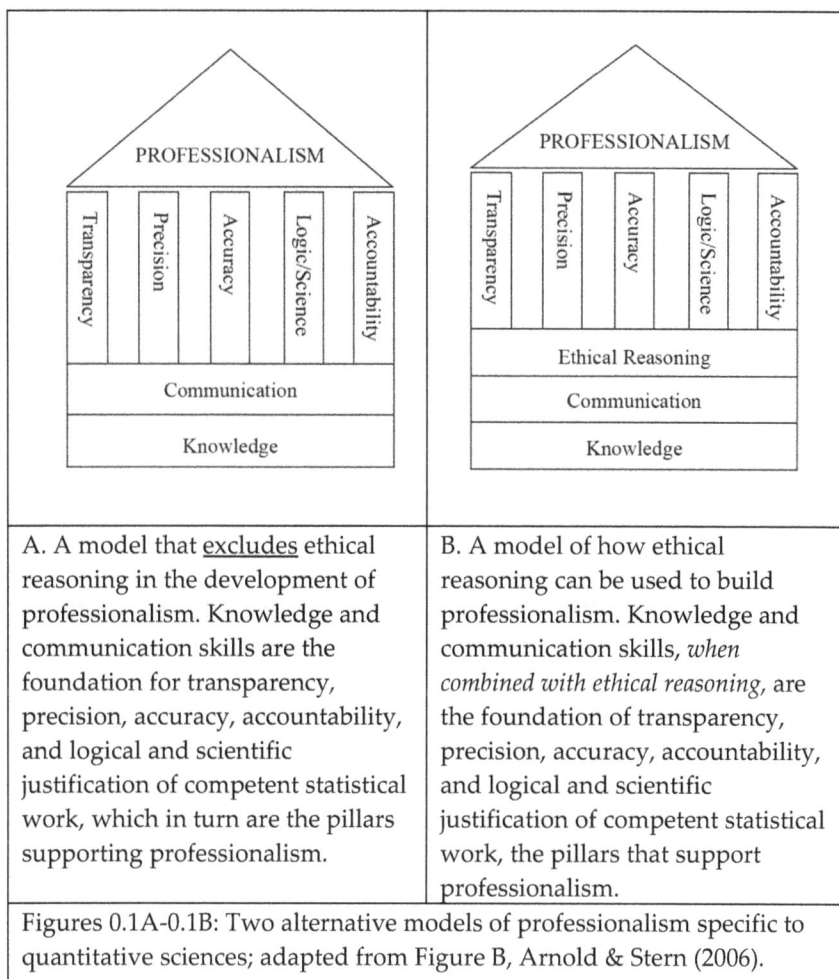

PROFESSIONALISM	PROFESSIONALISM
Transparency · Precision · Accuracy · Logic/Science · Accountability	Transparency · Precision · Accuracy · Logic/Science · Accountability
Communication	Ethical Reasoning
Knowledge	Communication
	Knowledge

A. A model that <u>excludes</u> ethical reasoning in the development of professionalism. Knowledge and communication skills are the foundation for transparency, precision, accuracy, accountability, and logical and scientific justification of competent statistical work, which in turn are the pillars supporting professionalism.	B. A model of how ethical reasoning can be used to build professionalism. Knowledge and communication skills, *when combined with ethical reasoning,* are the foundation of transparency, precision, accuracy, accountability, and logical and scientific justification of competent statistical work, the pillars that support professionalism.
Figures 0.1A-0.1B: Two alternative models of professionalism specific to quantitative sciences; adapted from Figure B, Arnold & Stern (2006).	

The models shown in Figures 0.1A and 0.1B support the conceptualization of ethical reasoning as integral for both effective work (e.g., Rosnow & Rosenthal, 2011) and the development of a sense of professionalism (e.g., Stern, 2006). For mathematics education, Ferrini-Mundy (2008) mentions "ethics" as an element of the core knowledge that PhD students in mathematics need, but does not discuss whether this can or should be integrated into this doctoral training as integral to the formation of the mathematician's professional identity. For undergraduate Data Science education, the National Academies reported that "...students will need exposure to material from multiple disciplines...and they will need training in ...ethical problem solving." This book is intended to provide such training, to prepare those who will use math, statistics, and data science to do so ethically in a data-centered world.

Organization of the book

The book is structured around engaging the reader in increasingly sophisticated engagement with ethical reasoning (ER) and the professional practice standards. First, familiarity is built with these standards and how they can create a decision-making framework that a quantitative practitioner can always utilize (Section 1). In Section 2, this engagement deepens to allow the reader to understand and gain practice in how the standards – current and future revisions - can be used to plan quantitative work, roughly sorted into six tasks, plus teamwork (for a total of seven tasks). In the final section, the deepest level of engagement involves the reader learning how to formulate and justify decisions about responses that might be warranted to ethical challenges that can arise (and have arisen!) in the data-centered world.

Section 1. Background and introduction

This Section provides the background information as an introduction to the book and its structure. Part of this information is Ethical Reasoning (ER), which comprises six elements of knowledge, skill, and ability (KSAs). Part of the background are the ethical practice standards of statistical and data science practice from the American Statistical Association (2022), and the Association of Computing Machinery (2018). Other support for ER includes the Stakeholder Analysis (SHA; Tractenberg, 2019-d), plus other key topics of interest for ethical engagement with data. Chapters 1.3-1.6 are all examples of "prerequisite knowledge", which is the first KSA of Ethical Reasoning (ER KSA 1), and its central importance in ethical reasoning. KSA 2 relates to choosing an ethical framework from which decisions can be made. Even before any ethical challenge or problem arises, a better understanding of why the GLs/CE exist can be gained by exploring two ethical frameworks that can be useful in decision making: "virtue ethics" and "utilitarian ethics". This prerequisite knowledge is summarized in Chapter 1.7, with additional information shared in Chapter 1.8. In Chapter 1.9, the reader moves on to ER KSAs 3, 4 (Chapter 1.10), with 5 and 6 presented together in Chapter 1.11.

Chapter 1.1. Introduction and Background

Introducing frameworks for ethical practice in statistics and data science

Introduce Bloom's taxonomy of cognitive behaviors (Bloom et al. 1956), explain its relevance for the book. *This Section is focused on Bloom's 1-4*

Chapter 1.2. Ethical Reasoning

Ethical Reasoning (ER) comprises six elements of knowledge, skill, and ability (KSAs). This chapter introduces ER and its importance, contrasting it with other "methods" for identifying, making decisions about, and justifying decisions about ethical challenges that arise whenever data are involved. Engagement with data goes beyond just data analysis –whether the individual analyzes the data formally themselves (e.g., as a statistician) or whether algorithms are used or created to analyze the data using software or other technology (e.g., as a data scientist).

Chapter 1.3. ASA Ethical Guidelines

This chapter presents the Guidelines (GLs) for Ethical Practice from the American Statistical Association (ASA, 2022).

Chapter 1.4. ACM Code of Ethics

This chapter presents the Association for Computing Machinery (ACM, 2018) Code of Ethics (CE).

Chapter 1.5. Prerequisite knowledge common to ACM CE and ASA GLs

This chapter presents the alignment – or concordance - of ASA and ACM Guidelines/Code, summarizing prerequisite knowledge that is common to both ACM CE and ASA GLs, and the "virtue ethics" framework for decision making. While not proposing that the virtue ethics model is better or worse than any other, understanding what this framework provides in terms of decision-making support is important for a full understanding of ethical reasoning.

Chapter 1.6. Stakeholder Analysis and Utilitarian Decision-making Framework

The Stakeholder Analysis and Utilitarian framework for decision making are presented. The utilitarian perspective may be a familiar one to readers, but how it can interact with professional practice standards will not be (because such standards do not typically integrate decision making frameworks). A key challenge in implementing the Utilitarian decision-making framework can be in the recognition and estimations of "harms" and "goods" or benefits, and the stakeholder analysis can help us to do this.

Chapter 1.7. Summarising KSAs 1 & 2

This chapter recaps the prerequisite knowledge and the KSAs of ER, focusing on prerequisite knowledge and decision-making frameworks, KSAs 1-2.

Chapter 1.8. Aligning prior training to promote ethical qualitative practice

This chapter discusses "other" ethics training readers in the US may have had, from the perspectives (and according to the priorities) of the US National Institutes of Health and National Science Foundations. These topics represent ancillary prerequisite knowledge, as well as definitions of human subjects, "research", and the concept of informed consent represent norms that are essential to understanding basic human rights (particularly of autonomy) and how these are addressed in the ethical practice standards of the ASA and ACM.

Chapter 1.9. Identify or recognize the ethical issue KSA 3

Returning to the ER KSAs, Chapter 1.9 discusses KSA 3, identifying or recognizing the ethical issue in any case or situation. This is one of the most difficult parts of the ER process, and is crucially dependent on a firm understanding of ethical practice standards like the GLs and CE, and also utilizes the Stakeholder Analysis (SHA).

Chapter 1.10. Identify alternative actions KSA 4

This chapter focuses on using the ethical practice standards to accomplish KSA 4, identify alternative actions (on the ethical issue), is only slightly less difficult that KSA 3. Two examples demonstrate how "do nothing" or "ignore the situation" are contrary to both the ACM and ASA standards, but offer a general approach to identifying plausible options for responding to an ethical issue (KSA 3) that are consistent with the standards.

Chapter 1.11. Make and justify a decision and reflect on it KSAs 5 & 6

The final ER KSAs, make and justify decision (KSA 5), and reflect on that decision (KSA 6) are discussed, in terms of how they represent (concretely) ethical practice, and also how they have the potential to influence "norms" in the workplace.

Section 2. Establishing familiarity with ASA and ACM principles/elements as they relate to the tasks (anticipating what problems may arise)

Introduction to Section 2: This section focuses the reader's development of cognitive capabilities on Bloom's Taxonomy levels 3-5, while reinforcing Bloom's 1-2 level familiarity with the practice standards. Emphasis is on understanding the relevance of GL and CE content for each of the tasks involved in work with data, leaning towards synthesis (of your experience with your new knowledge; or of different/diverse types of knowledge). Section 2 directs attention to how the SHA, GL, and CE can all provide guidance for your thinking/planning/workflows, as well as for each of the typical tasks in statistics and data science. Structure of the section is to introduce a SHA for each task. This raises the reader's Bloom's level (because SHA requires *analysis* (Bloom's 3) and *predictions* (Bloom's 4)) while reinforcing the reader's familiarity with the core content (Bloom's 1-2).

In each of the Section 2 chapters (2.2-2.9), both the ASA and ACM guidance on each task is identified. A SHA is completed, outlining harms and benefits that accrue when the task is completed with/without compliance with the ethical practice standards. Discussion questions are included in each chapter.

Chapter 2.1. Introduction to Section 2

Chapter 2.2. Planning/Designing

Chapter 2.3. Data collection/munging/wrangling

Chapter 2.4. Analysis (perform or program to perform)

Chapter 2.5. Interpretation

Chapter 2.6. Documenting your work

Chapter 2.7. Reporting your results/communication

Chapter 2.8. Engaging in team science/teamwork

Chapter 2.9. Summary of ASA and ACM Guidance on seven tasks

Section 3. Ethical reasoning using ASA and ACM principles/elements: vignettes

Chapter 3.1. Introduction to Section 3.

This section focuses the reader's development of Bloom's 4-6 thinking while reinforcing Bloom's 1-5 level engagement with the practice standards. In Section 3, there are ethical challenges that need identifying (ER KSA 3) and these come from an evaluation (Bloom's 5) of a brief vignette describing workplace events.

In this section the emphasis is on responding to the ethical challenge (i.e., using ER KSAs 3-6), requiring synthesis of prerequisite knowledge (GLs/CE) and understanding how the SHA drives the decisions as well as the justifications for how to respond to these workplace situations. The vignettes describe actual events relating to each of the typical tasks in statistics and data science (as introduced in Section 2). The vignettes are analyzed following the ER KSAs, including reflection that combines what a typical practitioner might consider about the case, and also thoughts about what the decision means from an instructional perspective. In this section, all KSAs of ethical reasoning are re-introduced and utilized/practiced.

Chapter 3.2. Planning/Designing

You are directed to design a system to scrape data from a specific source (e.g., Facebook), and are provided with specific design features of the source to ensure every data type can be scraped from every user.

Chapter 3.3. Data collection/munging/wrangling

You build a data scrape algorithm with a built-in opt-in feature, that will pop up and ask the user to opt-in to the data scraping (i.e., give consent) every time the algorithm changes, to scrape/collect more, or different data. That feature is removed.

Chapter 3.4. Analysis (perform or program to perform)

Piles of data start arriving from "the company data scraper" for you to analyze. No information is available that indicates whether consent was given for any of the data. You suspect, and then potentially identify an error in the code that removed that information.

Chapter 3.5. Interpretation

Results suggest that some people using the source (e.g., Facebook) are more susceptible to messaging (e.g., advertisements and fake news items) than others. You interpret this as signaling a need for caution/care in what your system does next – in order to limit bias, and support valid conclusions. Instead, you find reports of your work interpret the results without any caveats, and have removed any of your suggestions that sensitivity analyses may be needed.

Chapter 3.6. Documenting your work

You are told not to document your work. When you do (because that's what the practice standards say the ethical practitioner does), your boss returns it to you with the direction, "fix this".

Chapter 3.7. Reporting your results/communication

You submit your complete and correct report of your scraping algorithm – including identification of the removal of your built in, opt-in consent to contribute data; the lack of consent accompanying data to be analyzed; and the lack of your recommendations in interpretations for limiting bias. You later discover that none of this documentation was included in the final report, but the final report is shared with stakeholders as if it is complete and correct.

Chapter 3.8. Engaging in team science/teamwork

Leadership informs your team that they bought an algorithm that you will be using to scrape data. But first, they want you to take off all the consent pop-ups, because "that ruins the user experience" and "adds personal data we will only need to strip off to preserve confidentiality".

Chapter 3.9. Embracing your inner ethical practitioner

This chapter invites the reader to return to each of the chapters and vignettes in Section 3. Each case analysis that is provided is revisited through the lens of *role playing*. Specifically, readers are invited to practice – in actual role-playing dyads, or in writing, or both – the delivery of their ethically-reasoned decision in the workplace. Readers may feel more confident in engaging in conversations with others with a full, written analysis of what the issue is/issues are, what alternatives were considered, what decision is recommended/was taken and why, and some reflection on the relevance of the decision. This engagement is a specific part of ethical practice according to the

ACM and ASA, and is encouraged by the National Academy of Engineering (2013) and National Academies of Science, Engineering, and Medicine (NASEM, 2017). Initiating, and/or engaging in, that kind of conversation requires at least some practice! Readers should role play both sides of the conversation:

- describe your analysis of the case and your decision;
- "receive" a case analysis, and collaboratively determine the best way to ensure such situations do not recur;
- "receive" a case analysis and try to dissuade the analyzer from making – or publicizing – that decision;
- respond to someone who does not support your analysis and your decision.

Chapter 3.10. Summary of Section 3 and the book: career spanning engagement in professional and ethical practice

Reconsider what you have learned, and go on to refine what you have learned.

In each chapter of Section 3, we analyzed one vignette per task and readers were invited to analyze that case analysis. The objective of the Section was to teach and give practice in the full range of KSAs that are required for ethical reasoning – and not to make sure every conceivable situation was explored. The reader is encouraged to continue with self-directed learning, using the same KSAs on new problems and in particular, reflecting on the utility of each analysis to improve the chances of an *ethical* data-centered world. To support this self-direction, the reader is invited to consider, and discuss, how the ASA Guidelines and ACM Code of Ethics promote professionalism (in you) or the profession of statistics and data science (more generally). For example, consideration of the following:

- Explain whether/how the application of the Guidelines in any given case encourages ethical conduct in research (or practice, as appropriate) more generally.

- Do the ASA Guidelines and/or ACM Code of Ethics promote *professionalism*? How/how not?

Section 1. Background and introduction

Chapter 1.1

Introducing frameworks for ethical practice in statistics and data science[4]

1. Ethical practice = Frameworks + Reasoning

Data science is a new discipline, but it arises from two disciplines with long-standing commitments to ethical practice: computing and statistics. Ethical guidelines have been developed over several decades to support ethical professional practice with – as well as the application of – tools, techniques, and methods from both statistics and computing (Tractenberg et al. 2015; see also Tractenberg 2019-a). Engaging and practicing in the data-centered world requires some understanding of what it means to be an ethical statistician and an ethical data scientist.

Ethics is defined as "the moral principles that govern a person's behavior or the conducting of an activity". That definition seems straightforward, but there is a problem: the moral principles that govern *a person* might not be the same ones that govern *me*. Two refinements on this definition may help us: "normative ethics" is involved with what most people (in a given context) would consider to be "right" or "wrong" ways of behaving. The fact that "most people" feel something is right or wrong means that feeling (about rightness or wrongness) is what makes that characteristic (rightness/wrongness) a *norm*, or normal, general, typical behavior. Obviously, people who are new to a community where there <u>are</u> norms may find it difficult to deduce or infer what is right and wrong; and when there is no formal community at all, then there can be no norms. More specifically, since normative ethics concerns itself with describing behaviors as right or wrong, when new behaviors become part of (or even just available to) the community, there is a clear need to determine whether they are right or wrong.

---

[4] This chapter includes material that was originally published in Tractenberg RE. (2016). Why and How the ASA Ethical Guidelines should be integrated into every quantitative course. *Proceedings of the 2016 Joint Statistical Meetings, Chicago, IL.* Pp. 517-535; and in Tractenberg RE. (2020, February 19). Concordance of professional ethical practice standards for the domain of Data Science: A white paper. Published in the *Open Archive of the Social Sciences* (SocArXiv), 10.31235/osf.io/p7rj2

So far, we have a few problems: Firstly, "behaving morally" needs to be described such that all persons are, not just "a person" is, governed by the same principles. If this is not the case, then anyone can behave any way they prefer and call it "moral". Secondly, if rightness and wrongness are determined by norms, then "most people" need to be involved in the decisions about what behaviors fall onto the "Right" and "wrong" columns. When behaviors are determined to be "so wrong" that the community decides they should be punished, these tend to become barred by law (rather than simply by custom or norms). However, as you know/can imagine, something has to be described very precisely in order for it to be clear that a person has committed a crime or broken a law. Ethics tend not to be described so clearly – they do not rise to the level of "laws", even when they are true cultural or community norms. This brings us to the third problem: there needs to be fairly clear descriptions of both what is "right" and what is "wrong"; but, as it turns out, so many things are "right", and typically so few are "wrong", it tends to be easier to focus on what is "wrong" – so we can educate all newcomers to a community about what to avoid. While that is certainly easier, it does not give the newcomer a sense of "how to behave" like someone who contributes to those "norms" of right behaviors. For these reasons, and many others, communities of professionals have drafted ethical guidelines, or codes of ethics, which describe what "the ethical practitioner" (or more precisely, *how* the ethical practitioner) does their job. The implication is that if a practitioner does not follow the guidelines/code, then they are not doing their job ethically.

There is another type or branch of ethics, "applied ethics", which concerns itself with specific activities within well-defined communities (e.g., ethics applied to business is "business ethics"; ethics applied to biomedical sciences is "bioethics"). An important consideration for our purposes is, what if a quantitative practitioner sometimes works in business settings and other times in biomedical settings? Consultants may be described this way, because they are always quantitative practitioners, but the community in which they practice may change over time (this is also true for people who change jobs). Another situation is when a quantitative practitioner is actually working in *both* business *and* biomedical settings, like a biostatistician working for a pharmaceutical company. As you will see in later chapters, professional quantitatively oriented societies that have ethical guidelines (our example is the American Statistical Association, ASA) or a code of ethics (our example is the Association of Computing Machinery, ACM) state clearly that:

a. any practitioner who utilizes the tools/methods/techniques of the field are expected to follow *these* ethical guidelines/codes; and

b. any practitioner who is correctly following these guidelines/codes (i.e., because they are utilizing tools/methods/techniques of the field) should not allow people following other guidelines – other ethical norms – to cause them to *violate* the quantitative practitioner guidelines/codes.

Thus, these quantitative communities seek to ensure that their norms are followed whenever their tools/methods/techniques are employed. This is one way that these organizations – and communities – seek to promote *ethical* quantitative practice. These codes/guidelines are therefore standards for identifying "ethical practice" – as well as doing it, i.e., they are ethical practice standards. They outline "how to behave" like someone who contributes to "norms" of right behaviors in quantitative practices, and in general, when an individual fails to follow these norms, they are acting "wrongly". It is important to point out that "following the norms" should be observable – any two observers of the person acting should be able to agree, generally, on whether that was "right according to norms" or "wrong according to norms". If "acting wrongly" cannot be recognized sufficiently well to describe it to others (e.g., "I will know wrong behavior when I see it, but cannot describe it to you"), then it may not actually be sufficiently well-defined for everyone in the community to agree that it is, in fact, "wrong". This is another reason why guidelines are so important: only what is agreed on as "right" (or "wrong") are included in the guidelines, so people know fairly specifically what behaviors to avoid, and what behaviors mark the "ethical professional practitioner" (see Simonite 2018 for discussion of the relevance for this perception for "data science").

"Graduate instruction in statistics requires the presentation of general frameworks and how to reason from these." (Hubert & Wainer, 2011:62). This statement summarizes the perspective that teaching those who will be using statistics at work (graduate students across disciplines in their example) must acknowledge that the learners need to know more than just the formulas or how to use software. The learners also need what is sometimes referred to as "statistical literacy", or "statistical thinking", so that they can always identify the correct method, which exists within a general framework (e.g., categorical vs. continuous variables), and can also reason from the frameworks to the exact methodological features; and then reason about the results. The point Hubert & Wainer make is that "instruction in statistics" is not limited to just the statistical methodology: the frameworks *and* the ability to reason using those frameworks must both be learned. This also true for "instruction in ethics for quantitative practice": it requires the presentation of general *ethical* frameworks, and instruction and practice in how to reason from *these*. Professional societies have articulated guidelines that reflect the mindset of expert quantitative practitioners relating to their professional practice. These guidelines and codes are frameworks for ethical practice in the quantitative domains (Tractenberg, 2020).

Following the logic of Hubert & Wainer, though, providing the ethical guidelines or code of ethics to students or those new to the quantitative professions is not sufficient: "how to reason from these" is also a necessary component of ethics education. If you consider the Hubert & Wainer quote in the context of training for ethical quantitative practice, and you take the frameworks to be those professional practice standards, then "how to reason from these" is clearly the missing piece. "…(E)thics is not a vaccine that can be administered in one dose and have long lasting effects no matter how often, or in what conditions, the subject is exposed to the disease agent" (National Academy of Engineering, 2008 p. 36). Together, both the ethical frameworks *and* the ability to reason from these can shift even a single course in ethical practice from "one dose" that the National Academy of Engineering and National Research Council note is insufficient, towards the ability to reason ethically throughout a career, which is what the practice standards exist to promote.

This book was written to help familiarize the reader with these frameworks while also providing opportunities to practice reasoning from them. Some statements about ethical practice tend to –erroneously- reinforce the idea that ethical practice habits are simple to form (see Tractenberg, 2018). If this were true, it would mean that all of the unforeseen ethical challenges throughout statistics and data science were created purposefully and intentionally, because acting ethically is so simple and natural (that acting against those simple/natural instincts would require a lot of work on your part; Thiel 2015). While that might be a fair assumption for *illegal* behavior, it is not a reasonable conclusion about unethical behaviors.

Ethical challenges can arise in new and wholly unexpected situations throughout a career. Without training in how to reason with, or use, the ethical framework to prepare users of statistics for ongoing development of abilities to identify and reason through ethical challenges, it is not plausible to assume that these individuals will somehow prepare themselves. When "ethics training" focuses on static information or rules, the actual utility of that training is intrinsically limited. That is why this book focuses on ethical reasoning instead.

Every day and around the world, many individuals without professional statistician accreditation (PStat®) or even comprehensive training in statistics are asked to carry out statistical and data science tasks in both business and research settings –and this is an increasingly common situation as software, applications, and a perceived need for data analysis become ever more ubiquitous. It is untenable to assume that the training and practice working with ethical guidelines that are essential for developing the requisite familiarity with the GLs to support ethical statistical practice would be accomplished or even initiated by the single, *general-institutional* training module in "responsible conduct of research" that universities in the United States are required to provide for individuals receiving federal funding. This may be even less reasonable to assume for the institutional "ethics" training that many businesses and companies require of employees. Instead, if the ASA Ethical Guidelines and/or ACM Code of Ethics were introduced early, and reinforced throughout a curriculum to promote a sense of their relevance and ongoing engagement, this would result in long-term ethics education that would be obviously and specifically relevant to both students *and* faculty in the discipline of statistics. Moreover, trainees/students who are learning statistics or data science from or for *other disciplinary perspectives* would also learn both how to engage in these same important conversations about the ethical dimensions of statistical research and practice – *and also that such conversations are important.* Introducing ethical reasoning with/about data is important for improving the reproducibility of science across disciplines (see e.g. Freedman 2010; Collins & Tabak, 2014; McNutt 2014). For statistics and data science (encompassing practitioners in the disciplines of statistics, data science, and their intersection), attention to the features and standards of ethical practice will enrich both the career and the profession. This book exists to help with this enrichment.

2. Practice standards = Ethical Guidelines/Code of Ethics = Frameworks

The ASA Ethical Guidelines for Statistical Practice (American Statistical Association, 2022) comprise 8 core Principles plus an Appendix for organizations and institutions, which entail a total of 72 specific elements. The full (2022) Guidelines appear in following chapters, so here we just explore the topics:

A. Professional Integrity & Accountability (12)
B. Integrity of data and methods (7)
C. Responsibilities to Stakeholders (8)
D. Responsibilities to research subjects, data subjects, or those directly affected by statistical practices, Data Subjects, or those directly affected by statistical practices (11)
E. Responsibilities to members of multidisciplinary teams (4)
F. Responsibilities to Fellow Statistical Practitioners and the Profession (5)

G. Responsibilities of Leaders, Supervisors, and Mentors in Statistical Practice (5)

H. Responsibilities regarding potential misconduct (8)

APPENDIX: Responsibilities of organizations/institutions (12)

Meanwhile, the ACM Code of Ethics, updated in 2018, has four core areas, with 2-9 elements in each (ACM, 2018):

1. General Moral Principles (7)
2. Professional Responsibilities (9)
3. Professional Leadership Principles (7)
4. Compliance with the Code (2)

The 2022 Guidelines (ASA) and 2018 Code of Ethics (ACM) are presented in full, and discussed, in Chapters 1.3-1.5. These guidelines/codes are "the frameworks" that this book will help the reader learn to "reason from". At this point it is sufficient to notice the complexity and richness of these guidance documents. While certainly not the only such guidance, they are the professional standards for quantitative practitioners to be explored in this book. Because this book will teach the reader how to reason from *any such framework*, readers can go on to apply the reasoning to be presented to any other practice standard that is relevant (however, the practitioner should keep in mind that the ASA and ACM seek to promote ethical practice, and those communities hope that **their norms are followed whenever their tools/methods/techniques are employed**, in order to promote *ethical* quantitative practice). Other guidance may not be as comprehensive (e.g., International Statistics Institute, ISI, https://isi-web.org/index.php/activities/professional-ethics/isi-declaration has 12 principles[5]), but lack of a statement about something covered by ASA or ACM does not undermine the relevance or applicability of the ASA or ACM norms. This is primarily because these two organizations are among the largest – and therefore, most representative – for quantitative practitioners who are primarily statistical (ASA) or computational (ACM). The ISI ethical statement includes something that the ACM did not have in their 2018 version:

11. Bearing Responsibility for the Integrity of the Discipline
Statisticians are subject to the general moral rules of scientific and scholarly conduct: they should not deceive or knowingly misrepresent or attempt to

---

[5] International Statistical Institute (2010). Declaration on Professional Ethics (revised). Downloaded from http://www.isi-web.org/about-isi/professional-ethics/43-about/about/296-declarationprofessionalethics-2010uk on 11 Dec 2013.

prevent reporting of misconduct or obstruct the scientific/scholarly research of others.

The 2022 ASA Ethical Guidelines include several specific items that capture the same ideas (e.g., A1, B2-B3, C2, C8, D11, E4, H4-H5, H8). However, not all statistical practitioners, and data scientists, are doing scientific or scholarly work. Ethical reasoning, the 2018 ACM Code of Ethics, and the 2022 ASA Ethical Guidelines, are all specifically formulated for all practitioners. The guidance supports professional as well as scholarly and scientific applications of statistics, computing, and data science by practitioners at all levels.

Statisticians and data scientists, among other quantitative practitioners, have a special obligation to acknowledge and accept their responsibility for the integrity of the domain. In fact, Golde & Walker (2006) assert that, "Upon entry into practice, all professionals assume at least a tacit responsibility for the quality and integrity of their own work and that of colleagues. They also take on a responsibility to the larger public for the standards of practice associated with the profession." (p.10). This quote articulates an *assumption* about professional practice – namely, that everyone who enters a profession has been sufficiently prepared to take on the responsibilities for the standards that define that profession or discipline. The assertion was made (in the 2006 book on doctoral level preparation) in the context where individuals will have been formally trained as *stewards* of a particular discipline – those to whom the vigor, quality, and integrity of a particular discipline can be entrusted (Golde & Walker 2006: p. 5; see also Rios et al. 2019). The ASA and ACM do not (currently, in 2022) explicitly refer to disciplinary or professional stewardship, although the 2022 ASA Guidelines do specify (Principle F) that practitioners have responsibilities to both other practitioners and to the profession itself. When quantitative practitioners do not follow ISI statement 11, or the general practice standards of the ASA or ACM, then all those who make decisions on the basis of the results of quantitative practice may find their decisions, or their scientific or scholarly work, undermined.

Tractenberg (2016-a; 2016-b) discusses why training with the ASA Ethical Guidelines is important for training graduate and undergraduates in the quantitative sciences – these reasons pertain to the ACM Code of Ethics as well (see also Tractenberg, 2020). Four objectives or rationales for integrating training in ethical practice are to:

A. **Encourage ethical conduct in (throughout) the practice of statistics and data science**, by pointing out how everyone on a team has their specific role with its attendant obligations and priorities (i.e., to bring ISI principle 11 to *all* quantitative practice).

B. **Promote professionalism for all of the team members**, irrespective of their level of training in statistics and/or data science.

C. **Promote the consideration, prior to the start of analyses, of the analyses and the qualifications of the analyst/data scientist** to plan, execute, and interpret them.

D. **Engage with principles of professional practice for statisticians and data scientists**, which can promote both appreciation for the statistician and data scientist as a collaborating team member and understanding how this team member is accountable and responsible for their work.

Professional statisticians and data scientists are responsible for being ethical in their practice. However, responsibility for ethical behavior goes beyond the individual practitioner: both the ASA and ACM state that their practice standards are intended to promote ethical behaviors for both those who self-identify as their specific target audience (statisticians and ACM members) but are also applicable for *all those who utilize their technology/methods and approaches*. Further, both ASA and ACM practice standards have specific sections that address the responsibilities that employers (ASA) or leaders (ACM) have. It is impossible to require or even suggest that the statistician or data scientist is responsible for ensuring either that their attention to ethical practice is shared/honored by the rest of the team/their colleagues or that all participants in a project or workplace follow the ASA/ACM practice standards. However, when a practitioner encounters resistance or antagonism as they endeavor to practice ethically, it is evidence of violations of ACM Principle 3 (promoting ethical leadership in the workplace) and ASA Principle G (leading statistical practitioners so as to promote ethical behavior and an ethical workplace in which the statistician or data scientist feels comfortable).

It may seem like an unfair burden, but in fact, doing what you can to promote an ethical workplace strengthens not only your own practice but also that of your colleagues. This is not actually an expectation unique to statistical and quantitative practice; as noted, Golde & Walker assume this of all those who enter any professional context! However, statistics and data science are perhaps unique in the specificity of their ethical practice standards – which allows quantitative practitioners in these fields to actually follow the standards, and meet this expectation. One of the best ways to encourage real and positive attention to ethical practice in the workplace is to refuse to work in environments where ethical practice is not a priority or where it is actively resisted or discouraged. In some cases, the workflow is so distributed that there appears to be no single individual with the specific responsibility for ethical treatment of data, its analysis, or how the results are reported, maintained, and evaluated to ensure that harms do not accrue. Knowing what you know (or will, after you read this book!) about the responsibilities of those who work with data according to the ASA and ACM, recognizing that diffusion has effectively removed considerations of responsibilities for ethical practice would be an excellent reason to turn down, or not even seek, a position in such a work environment.

These practice standards are as important – and useful! – "for non-professional" data analysts and data scientists as they are for accredited and professional statisticians, computing practitioners, and data scientists. Moreover, these standards are also essential for leaders, supervisors, and employers who work with practitioners in the field. "Ethical principles are worse than useless if we don't allow them to change our practice, if they don't have any effect on what we do day-to-day." (Loukides et al. 2018). The purpose of this book is to support your use – every day, as needed - of the ethical practice standards outlined by the ASA and ACM. The book is focused on the most recent versions of these practice standards, but the reasoning approach can be used with older and newer versions – as well as with other ethical practice standards and guidelines, or workplace policies such as Principles and Practices for a Federal Statistical Agency (National Academies of Sciences, Engineering, and Medicine, 2021), the US Government Statistics Agency ethical practice standard.

3. Reasoning

The number of specific elements suggest that memorizing the ASA (72) or ACM (25) Guidelines/Code is an unwieldy task. Both ASA and ACM preambles state that their practice standards offer *guidance*, not rules; and that each practitioner – while responsible for following their disciplinary guidelines and following local rules/laws – must also utilize their judgement in order to practice ethically. That is, the abilities to effectively utilize the ethical practice standards will require a degree of ethical reasoning: "Ethics is the effort to guide one's conduct with careful reasoning. One cannot simply claim "X is wrong."; Rather, one needs to claim, "X is wrong because (fill in the blank)"." (Briggle & Mitcham, 2012, p. 38; see also Tractenberg & FitzGerald, 2012). As you are 'filling in the blank' that Briggle & Mitcham identified, you can also deepen your understanding, and gain practice with application, of ASA and ACM ethical practice principles. More to the point, however, "reasoning" is much more than "memorizing". In order to encourage the ethical use of statistics and data science, everyone learning use these methods/techniques also needs to learn how to utilize judgement, rather than memorization, as they learn about what constitutes *ethical* use of these disciplines, instructors and self-directed learners can leverage Bloom's taxonomy of cognitive behaviors (Bloom et al. 1956). We appeal to Bloom's Taxonomy of Educational Objectives for the Cognitive Domain (Bloom et al. 1956). This is a six-level hierarchy of cognitive behaviors (read more about it on Wikipedia, https://en.wikipedia.org/ wiki/Bloom%27s_taxonomy). The following table was based on Bloom's but includes elaboration in terms of all the different sorts of thinking that you might be familiar with having done/having to do throughout your educational experience. This table shows the different cognitive behaviors ranging from least complex (remember/reiterate) to most complex (create/apply). The fact that this is a *hierarchy* means that it is important to be able to successfully do earlier level behaviors before you can move on to then be able to engage in more complex behaviors – even with the same material or purpose. The table can help demonstrate how different "memorization" (the first level – least cognitively demanding and complex) is from "reasoning" (the highest level – most cognitively demanding and most complex).

a. Remember/Reiterate-answer based on recognition of previously seen example Locate or retrieve relevant knowledge from memory. Recognize, reproduce, recall, restate (verbatim), apply labels (simple recall)/fill in the blank.
b. Understand/Summarize- answer summarizes info already in question (and/or answers) Report/summarize, focused recall, discriminate relevant from irrelevant info. Change from one type of representation into another (e.g., paraphrase). Find a specific example (focused paraphrasing) or illustration of a concept or principle; matching.
c. Apply/ Illustrate-answer extrapolates from seen examples to (really) new examples Recognize previously unseen examples/exemplars; give a previously unseen example, **identify what examples represent**. Apply a procedure to a familiar or unfamiliar task.
d. Analyze/Predict Employ a rule to predict changes that are proximal or distal to the site of change or the time of change. Use given criteria to articulate what comes next- **separation of time/space differentiates this from simple selection and application of correct formula (which is at c level).** Determine how parts of a structure relate to each other and the whole; determine purpose.
e. Evaluate Use criteria to make a decision, judgment or selection; determine what criteria were used in making a judgment; analyze a situation or problem to determine the consequences; discuss both sides of an issue. Explain how a decision was reached. Detect inconsistencies or fallacies; determine efficiency and appropriateness of a procedure.
f. Create/Apply The opposite of prediction: describe the chain of events leading to an outcome; given an outcome, articulate the stages leading to the outcome (given the initial state). Problem solving; devise a procedure to accomplish a task; formulate an alternate hypothesis or course of action (rather than choose). Think and argue critically.

Table 1.1.1 *Cognitive Complexity Matrix (CCM)*[6]

The cognitive requirements to meet Briggle & Mitcham's definition of "careful reasoning", go beyond "memorization", as shown in the figure below (taken from Tractenberg, 2020).

---

[6] Table 1.1.1 adapted from Tractenberg RE, Gushta MM, Mulroney SE, Weissinger PA. (2013). Multiple choice questions can be designed or revised to challenge learners' critical thinking. *Advances in Health Sciences Education*, 18(5):945-61.

Bloom's Taxonomy of Cognitive Complexity and effective use of ethical practice standards

Individual is self-reflective and can iteratively reason using guidance — Bloom's Level 6 (**B6**): judgment & evaluation of rationale/justifications

Bloom's Level 5 (**B5**): synthesis (creating rationale)

Sufficient sophistication to utilize guidance for decision making — Bloom's Level 4 (**B4**): analysis, selection/justification of choices ·········· *minimum level of reasoning capability required to utilize practice standards effectively*

Basic knowledge of guidance and aspects of reasoning are in place — Bloom's Level 3 (**B3**): application — **B3**: **identify** which guidelines may be relevant to apply/whether checklist items are relevant

Bloom's Level 2 (**B2**): understanding — **B2**: **use** a checklist (yes/no) uncritically

Bloom's Level 1 (**B1**): memorization — **B1**: **recite** Ethical Practice Guidelines or other standard

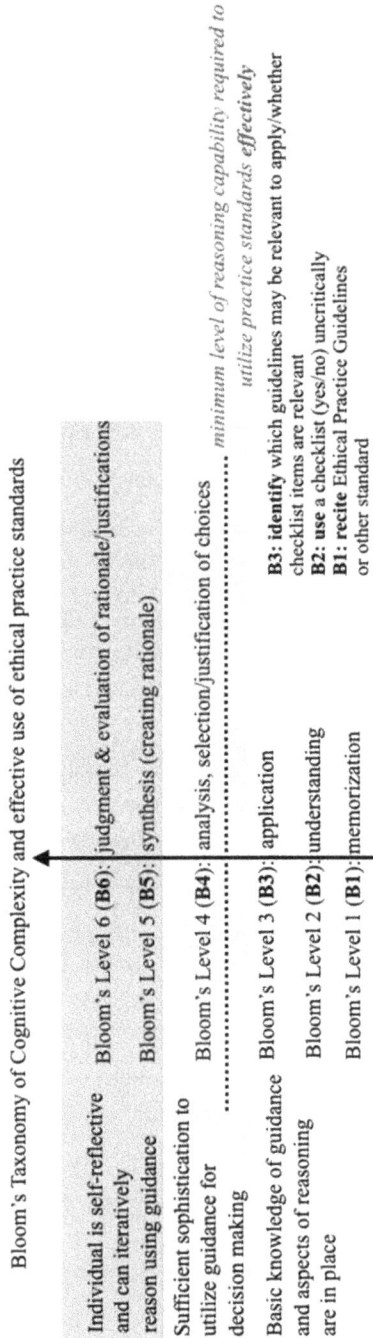

Figure 1.1.1. *Bloom's cognitive complexity level 4 ("Bloom's 4") is the minimum level required to apply the judgment needed to utilize ethical practice standards.*

As defined by Briggle & Mitcham (2012), the ability to explain/justify why one or another ethical guideline or principle pertains, or should be prioritized over other considerations, constitutes the target level of engagement with what constitutes "ethical practice of statistics and data science"; handing out (or testing on memorization of) the ethical guidelines themselves is clearly not sufficient.

Through the sections of this book, the reader will practice and focus first on Bloom's levels 1-3 (levels a-c in the CCM table: remember; understand; illustrate/apply) in Section 1, then in Section 2, the reader will practice levels a-c but also spend more time applying (level c in the table, Bloom's level 3) and analyzing (level d in the table, Bloom's level 4). In Section 3, Bloom's levels 1-6 will be used and reinforced, as ethical reasoning with vignettes (mini-cases) is explored. That is where we begin to evaluate (level e in the table, Bloom's level 5) and really "think and argue critically" (level f in the table, Bloom's level 6). All the verbs, from "reiterate" to "judge", are methods by which a learner processes - learns - material. The deeper/more extensive the processing, the more readily available the information (Tulving et al. 1994); and this taxonomy is truly hierarchical.

This focus on cognitive complexity may feel like overkill, but this book is intended to teach ethical **reasoning**, and because reasoning requires a hierarchical developmental trajectory, that is how the book is structured. While practitioners need to know and understand the core principles (i.e., Bloom's 1/level a of the matrix), they also need to anticipate that, and recognize when, elements within one set of practice standards can sometimes be in conflict within a single case or situation. This highlights the earlier point that the framework is necessary but not sufficient for ethical practice. The ability to *reason* with/from the framework is also required. Reasoning from frameworks requires a great deal more than memorizing, and the intention is to help you grow these capabilities as you read so that you can continue to improve your higher Bloom's capabilities in new situations in the data-centered world.

Even if everyone practicing in the data-centered world was committed to memorizing the practice standards, simply distributing the standards will *not* address any of the four objectives of integrating ethical practice standards into professional preparation, as is shown in Table 1.1.2. Thus, learning to recognize these frameworks and how to reason with, and prioritize, their core principles and their constituent elements are advocated for every person who engages with data.

OBJECTIVES FOR INTEGRATING GUIDELINES: OPTIONS FOR INTEGRATING ASA ETHICAL GUIDELINES:	Encouraging ethical conduct throughout the practice of science	Promoting professionalism for all participants in research/data analysis	Promoting consideration of analysis features/ requirements	Engaging with principles of professional statistical practice
Direct students to the Guidelines website (ASA)				
Attach Guidelines to syllabus				
Discuss in one class meeting				x*
Integrate into existing course (i.e., discussion in at least 1/3 of meetings and assignments that are discussed in class during (some part) of 1/3 of meetings)			x	
Create stand-alone course	x	x	x	x
Integrate across courses in sequence/series (i.e., discussion in at least 1/3 of meetings and assignments that are discussed in class during (some part) of 1/3 of meetings –for each course)	xx	xx	xx	xx
Integrate through-out curriculum	xxx	xxx	xxx	xxx

Table 1.1.2: *Options for integrating ASA Ethical Guidelines- and whether they achieve any teaching or learning objectives for doing so.*[7]

[7] Table reprinted from (Table 1) Tractenberg RE. (2016-c). Why and How the ASA Ethical Guidelines should be integrated into every quantitative course. *Proceedings of the 2016 Joint Statistical Meetings, Chicago, IL.* Pp. 517-535.

"The entire community of scientists and engineers benefits from diverse, ongoing options to engage in conversations about the ethical dimensions of research and (practice)," (Kalichman, 2013: 13). Simply memorizing the practice standards cannot support engagement in these important conversations about the ethical dimensions of your work. Preparing quantitative practitioners to engage competently in conversations about ethical practice requires purposeful, widespread, and developmental training that can come from, and support, a culture of ethical research and practice. As suggested by the National Academy of Engineering and National Research Council, whether or not statistics and data science will be your primary focus, training with ethical frameworks, plus learning to reason from these, should have "long lasting effects no matter how often, or in what conditions" you need to use them. In order to engage in reasoning (Briggle & Mitcham, 2012) or conversations (as suggested by Kalichman, 2013), greater depth of learning is required than simply remembering, which is the least complex level of processing according to Bloom's taxonomy.

4. Reasoning is not enough: *Ethical Reasoning* is needed

There is a specific type of reasoning from ethical frameworks: ethical reasoning. As you might imagine, if learning to reason requires you to develop a skillset leading up to the most complex Bloom's/cognitive behavior, then learning to reason with material that can seem to be more subjective – ethics –requires development of that whole skillset but under conditions of much more uncertainty than for other types of material. Recognizing this, a model was published in 2012 (Tractenberg & FitzGerald, 2012) describing purposeful engagement in the development and growth of a set of six learnable, improvable types of knowledge, skills or abilities (KSAs) that are the building blocks of ethical reasoning: Prerequisite knowledge; identification of decision-making frameworks; recognizing an ethical issue; identification and evaluation of alternative actions; making and justifying decisions; and reflecting on the decision. In that 2012 paper, "recognizing an ethical issue" was listed 2nd but this has been updated here, putting that third.

This book discusses ethical reasoning extensively in Chapter 2, and then Sections 2 and 3; here we simply point out the dependence of the "learnable and improvable" aspects of ethical reasoning on Bloom's Taxonomy.

The list of ethical reasoning KSAs is focused on decision-making and reasoning, and this is featured in the next chapter. We have argued elsewhere (Tractenberg & FitzGerald, 2015; Tractenberg et al. 2015; Tractenberg, 2016-b) that all data analysis requires decision-making, and whether or not an individual self-identifies as a "statistician", quantitative analysis –even if it derives from automation or algorithmic pattern matching – also requires decision-making. This is one reason why ethical reasoning is so crucial for the training of all quantitative scientists. The approach to ethical reasoning has been shown to lead to sustained learning Tractenberg, et al, 2017) and is also useful across diverse situations that a statistician might encounter (e.g., Gunaratna & Tractenberg, 2016). It is hoped that by presenting the approach with both the ASA and ACM practice standards in this book, you will also find it to be useful across data science contexts. This is explored in Sections 2 and 3.

If one accepts that "(g)raduate instruction in statistics requires the presentation of general frameworks and how to reason from these." (Hubert & Wainer, 2011:62), it is clear that*all* **instruction in statistics and quantitative science** *also* **requires the presentation of general** *ethical* **frameworks, and instruction and practice in how to reason from** *these*. Promoting this two-part view of ethics education in the preparation of statisticians and quantitative scientists for ethical practice should lead to "long lasting effects" and a common culture of ethical research and practice. This is a desirable outcome in general, and this book seeks to boost your awareness and understanding of the ethical considerations of everyday practice of statistics and data science–and to teach and give practice in the specific skills needed to identify these ethical considerations, and reason your way to a satisfactory *and defensible* decision.

Chapter 1.2

Ethical Reasoning –
Learnable, Improvable Knowledge,
Skills, and Abilities (KSAs)

As we saw in the previous chapter, "reasoning" is a more complex type of thinking than "remembering" in general, and reasoning about ethics, or reasoning to support ethical decision making, may seem more complex because "ethics" can seem to be less precise, or less objective, than other types of information that you might already have learned to reason with. This chapter describes ethical reasoning as a process that can be learned and improved.

We conceptualize "ethics education" as an ongoing, purposeful, engagement in the development and growth of a set of six learnable, improvable types of knowledge, skills or abilities (KSAs) that make up ethical reasoning. These six KSAs[8] of ethical reasoning are:

1. Determining your prerequisite knowledge;
2. Identification of decision-making frameworks;
3. Recognizing an ethical issue;
4. Identification and evaluation of alternative actions;
5. Making and justifying decisions; and
6. Reflecting on the decision (Tractenberg & FitzGerald, 2012; Tractenberg, et al, 2017; see also Gunaratna & Tractenberg, 2016).

This ethical reasoning KSA list is focused on decision-making and reasoning - and not on the mastery of information alone. We have argued elsewhere (Tractenberg & FitzGerald, 2015; Tractenberg et al. 2015; Tractenberg, 2016-b) that all data analysis requires decision-making, and whether or not an individual self-identifies as a "statistician". Work that involves data –even if it emphasizes automation, mining, machine learning or other algorithms– also requires decision-making. This is one reason why ethical reasoning is so crucial

---

[8] Note that the original list has KSAs 2 and 3 reversed. The KSAs are the same, but their order has been updated since Tractenberg & FitzGerald (2012). Identifying the ethical problem (originally #2) should happen *after* the decision-making framework was identified. Otherwise, it seems like you choose the framework that suits the problem – easily manipulated to become "choose the framework that yields the solution you prefer", which is inappropriate and unethical.

for the training of *all* quantitative practitioners. Preparing all scientists to engage competently in these conversations requires purposeful, widespread, and developmental training that can come from, and support, a culture of ethical research and practice.

Learning these KSAs does require – and promote – ongoing conversations about ethical dimensions of quantitative practice in research and other contexts. However, based on your reading of the previous chapter, you know that part of your prerequisite knowledge (KSA 1) will be in the form of the practice standards – meaning that some level of memorization – familiarity-with both ASA and ACM guidelines (see next three chapters) will be necessary. There are two other tools that can be helpful as we prepare to make decisions: the first we will study after the practice standards: a template for considering harms and benefits that can accrue, based on or due to decisions that are made. These harms and benefits accrue to different individuals, called "stakeholders" – because they have a stake, implicitly or explicitly, in the outcome of a decision or action. So, we will explore the stakeholder analysis, learning how to do this and also how the learning of how to assess the harms and benefits of our work will itself helps us to be more ethical overall. The final tool we will explore is a Data Science Ethics Checklist (DSEC). This tool was developed to support decision making in data science projects, and includes 20 questions, grouped into five areas: A. Data Collection; B. Data Storage; C. Analysis; D. Modeling; and E. Deployment. Unlike the practice standards discussed in the next three chapters, the DSEC does not discuss whether the practitioner has a special responsibility or obligation to answer all questions in the affirmative; as noted in the first chapter, the practice standards do state (ASA) or imply (ACM) that the practitioner has responsibilities to multiple stakeholders -that is, that the ethical practitioner can be recognized as the professional who accepts the responsibility, and acts to uphold it.

Tractenberg & FitzGerald (2012) describe ER as a series of steps but in fact, some of the steps involve other (implicit) steps, and the results of some of the steps may require revisiting or revising the output of earlier steps – in that way, the ER steps can be considered *iterative*. This chapter outlines all six KSAs, and the next three chapters focus on KSA 1, determining the extent of your knowledge about a given case, task, or situation to ensure that you will be able to reason about it; and KSA 2, identifying decision-making frameworks. (In fact, KSAs 1-2 are the focus of both Sections 1 and 2 – they're *that* important!)

The ethical reasoning KSAs can be elaborated:

1. **Identify and 'quantify' prerequisite knowledge**: this should include the ASA and ACM practice standards, plus other contextual information (e.g., stakeholder analysis, DSEC).

2. **Identify decision-making frameworks.** The virtue and utilitarianism frameworks will be discussed at length in the next two chapters. This step will consist of considering whether they suggest the same or different courses of action for the identified issue. More specifically, Guideline principles themselves may suggest different options and these must be recognized so that alternative actions can be formulated and evaluated.

3. **Identify or recognize the ethical issue.** The vignettes in the book generally describe "something wrong" but are not specific about what is actually wrong. Using the stakeholders' analysis results and the Guidelines, at least one ethical issue can be identified. If there are other ethical issues that a vignette raises, the most superficial/obvious one will be analyzed so that instructors can use the same vignette to allow students to do their own analyses (resulting from "the other" ethical issue(s)).

4. **Identify and evaluate alternative actions (on the ethical issue).** There are **always** three decisions that can be made in any circumstance:
    a) do nothing.
    b) consult or confer with a peer or a supervisor – using the professional guidelines or other resources.
    c) report violations of policy, procedure, ethical guidelines, or law.
    Alternatively, "agree to do <requested actions>", "do not agree to do <requested actions>"; or "ignore <request>".
    Every analysis in this book will include considerations of what "do nothing"/"ignore request" means for the stakeholders, and at least one more specific (than do nothing/ignore request) alternative actions will be considered. Coming up with that "other" alternative is one of the most challenging parts of ethical reasoning.

5. **Make and justify a decision.** The decision will be articulated, and justification will be based on the ASA Guidelines/ACM Code, and effects of the decision on stakeholders. So, while it is of course our objective to become familiar with these practice standards and the effects our decisions and actions can have on stakeholders, the fact that decisions must be *both made and justified* underlines our earlier assertion that "instruction in ethics for quantitative practice requires the presentation of general ethical frameworks, and instruction and

practice in how to reason from these." Also, the justification of a decision requires the highest level cognitive behaviors.

6.   **Reflect on the decision**. Consider your analysis, and determine if it yields the decision you feel most comfortable with – and can justify most strongly. If there are what appear to be equally justifiable alternatives, consider whether one might be more justifiable in some circumstances or for some stakeholders. It may also be the case that in the future, additional stakeholders may be identified, or other changes to the situation (or your prerequisite knowledge) mean the decision needs to be re-examined. Reflection can help notify others in the profession or in similar situations of the existence of the ethical issue that was encountered; it will also set the next "decider" up to do this decision-making process in a stronger way (because sources of prerequisite knowledge were already assembled, and can be reviewed/revised as appropriate). Reflection is also a high-level cognitive behavior.

In addition to reflecting on the process by which your decision was made and justified, and considering how the effort of making the decision using the ER KSAs could be made more efficient for others, or making the decision and its justification known to others might make similar ethical problems less likely to occur/more likely to be effectively dealt with, there are other reflection questions that can be considered:

- Why is "do nothing" / "ignore request" a decision that you can make? Is "do nothing" a justifiable decision? How so/why not?
- Why should you do or say something (in a case like this) when no one else seems to think there is a problem? Why or why not?
- What would you do if you encountered an ethical problem and, as you sought additional prerequisite knowledge, you consulted your supervisor, and they told you that "this happens all the time"? or that "Yes, that happens to me a lot"? Would you think the ER process, and particularly the inclusion of "reflection", was more or less important for quantitative practitioners? Why?
- Is "no one else seems to have a problem (with this)" ever a good answer? Why or why not?
- What justification is there for a claim that "ethical practice is just common sense applied at work"? Is that true (in this case)?
- Why do people say that there is "no right answer" to problems/ questions of ethical practice? Do you agree?

- Explain whether/how the application of the Guidelines/Code in this case encourages ethical conduct in research (or practice, as appropriate)
- Does consulting the ASA Guidelines or ACM Code of Ethics promote *professionalism* in you when applied to this case? How/how not?

Each of these six KSAs are discussed and practiced in Section 3. You may have noticed that all of the KSAs 3-6 actually rely on KSAs 1-2 ("Identify and 'quantify' prerequisite knowledge" (KSA 1) and "Identify decision-making frameworks" (KSA 2)) and the work done for these, which is why these are the main focus of Sections 1 and 2. However, we will re-visit KSAs 3-6 before focusing on KSAs 1-2 again in Section 2. You are encouraged to keep these reflection questions in mind as we move through the book. If your answers change (even if they change from "hm, I really don't know how to answer this!" to "I definitely do not know the answer, but I do understand the question", it will represent a lot of work on your part to "learn and improve" your KSAs 1 and 2. Since we recognize that KSAs 3-6 will depend almost entirely on KSAs 1-2, it makes some sense to focus on the prerequisite knowledge (KSA 1) and decision-making frameworks (KSA 2) so intently.

Chapter 1.3
ASA Ethical Guidelines for Statistical Practice

The ASA Ethical Guidelines for Statistical practice were prepared by the Committee on Professional Ethics and approved by the Board of Directors of the American Statistical Association in January 2022. The Guidelines appear here verbatim, and are also available at https://www.amstat.org/asa/files/pdfs/EthicalGuidelines.pdf

Purpose of the guidelines

The American Statistical Association's Ethical Guidelines for Statistical Practice are intended to help statistical practitioners make decisions ethically. In these Guidelines, "statistical practice" includes activities such as: designing the collection of, summarizing, processing, analyzing, interpreting, or presenting, data; as well as model or algorithm development and deployment. Throughout these Guidelines, the term "statistical practitioner" includes all those who engage in statistical practice, regardless of job title, profession, level, or field of degree. The Guidelines are intended for individuals, but these principles are also relevant to organizations that engage in statistical practice.

The Ethical Guidelines aim to promote accountability by informing those who rely on any aspects of statistical practice of the standards that they should expect. Society benefits from informed judgments supported by ethical statistical practice. All statistical practitioners are expected to follow these Guidelines and to encourage others to do the same.

In some situations, Guideline principles may require balancing of competing interests. If an unexpected ethical challenge arises, the ethical practitioner seeks guidance, not exceptions, in the Guidelines. To justify unethical behaviors, or to exploit gaps in the Guidelines, is unprofessional, and inconsistent with these Guidelines.

Principle A: Professional Integrity and Accountability

Professional integrity and accountability require taking responsibility for one's work. Ethical statistical practice supports valid and prudent decision making with appropriate methodology. The ethical statistical practitioner represents their capabilities and activities honestly, and treats others with respect.

The ethical statistical practitioner:

1. Takes responsibility for evaluating potential tasks, assessing whether they have (or can attain) sufficient competence to execute each task, and that the work and timeline are feasible. Does not solicit or deliver work for which they are not qualified, or that they would not be willing to have peer reviewed.
2. Uses methodology and data that are valid, relevant, and appropriate, without favoritism or prejudice, and in a manner intended to produce valid, interpretable, and reproducible results.
3. Does not knowingly conduct statistical practices that exploit vulnerable populations or create or perpetuate unfair outcomes.
4. Opposes efforts to predetermine or influence the results of statistical practices, and resists pressure to selectively interpret data.
5. Accepts full responsibility for their own work; does not take credit for the work of others; and gives credit to those who contribute. Respects and acknowledges the intellectual property of others.
6. Strives to follow, and encourages all collaborators to follow, an established protocol for authorship. Advocates for recognition commensurate with each person's contribution to the work. Recognizes that inclusion as an author does imply, while acknowledgement may imply, endorsement of the work.
7. Discloses conflicts of interest, financial and otherwise, and manages or resolves them according to established policies, regulations, and laws.
8. Promotes the dignity and fair treatment of all people. Neither engages in nor condones discrimination based on personal characteristics. Respects personal boundaries in interactions and avoids harassment including sexual harassment, bullying, and other abuses of power or authority.
9. Takes appropriate action when aware of deviations from these Guidelines by others.
10. Acquires and maintains competence through upgrading of skills as needed to maintain a high standard of practice.
11. Follows applicable policies, regulations, and laws relating to their professional work, unless there is a compelling ethical justification to do otherwise.
12. Upholds, respects, and promotes these Guidelines. Those who teach, train, or mentor in statistical practice have a special obligation to promote behavior that is consistent with these Guidelines.

Principle B: Integrity of Data and Methods

The ethical statistical practitioner seeks to understand and mitigate known or suspected limitations, defects, or biases in the data or methods and communicates potential impacts on the interpretation, conclusions, recommendations, decisions, or other results of statistical practices.

The ethical statistical practitioner:

1. Communicates data sources and fitness for use, including data generation and collection processes and known biases. Discloses and manages any conflicts of interest relating to the data sources. Communicates data processing and transformation procedures, including missing data handling.

2. Is transparent about assumptions made in the execution and interpretation of statistical practices including methods used, limitations, possible sources of error, and algorithmic biases. Conveys results or applications of statistical practices in ways that are honest and meaningful.

3. Communicates the stated purpose and the intended use of statistical practices. Is transparent regarding a priori versus post hoc objectives and planned versus unplanned statistical practices. Discloses when multiple comparisons are conducted, and any relevant adjustments.

4. Meets obligations to share the data used in the statistical practices, for example, for peer review and replication, as allowable. Respects expectations of data contributors when using or sharing data. Exercises due caution to protect proprietary and confidential data, including all data that might inappropriately harm data subjects.

5. Strives to promptly correct substantive errors discovered after publication or implementation. As appropriate, disseminates the correction publicly and/or to others relying on the results.

6. For models and algorithms designed to inform or implement decisions repeatedly, develops and/or implements plans to validate assumptions and assess performance over time, as needed. Considers criteria and mitigation plans for model or algorithm failure and retirement.

7. Explores and describes the effect of variation in human characteristics and groups on statistical practice when feasible and relevant.

Principle C: Responsibilities to Stakeholders

Those who fund, contribute to, use, or are affected by statistical practices are considered stakeholders. The ethical statistical practitioner respects the interests of stakeholders while practicing in compliance with these Guidelines.

The ethical statistical practitioner:

1. Seeks to establish what stakeholders hope to obtain from any specific project. Strives to obtain sufficient subject-matter knowledge to conduct meaningful and relevant statistical practice.
2. Regardless of personal or institutional interests or external pressures, does not use statistical practices to mislead any stakeholder.
3. Uses practices appropriate to exploratory and confirmatory phases of a project, differentiating findings from each so the stakeholders can understand and apply the results.
4. Informs stakeholders of the potential limitations on use and re-use of statistical practices in different contexts and offers guidance and alternatives, where appropriate, about scope, cost, and precision considerations that affect the utility of the statistical practice.
5. Explains any expected adverse consequences from failing to follow through on an agreed-upon sampling or analytic plan.
6. Strives to make new methodological knowledge widely available to provide benefits to society at large. Presents relevant findings, when possible, to advance public knowledge.
7. Understands and conforms to confidentiality requirements for data collection, release, and dissemination and any restrictions on its use established by the data provider (to the extent legally required). Protects the use and disclosure of data accordingly. Safeguards privileged information of the employer, client, or funder.
8. Prioritizes both scientific integrity and the principles outlined in these Guidelines when interests are in conflict.

Principle D: Responsibilities to research subjects, data subjects, or those directly affected by statistical practices

The ethical statistical practitioner does not misuse or condone the misuse of data. They protect and respect the rights and interests of human and animal subjects. These responsibilities extend to those who will be directly affected by statistical practices.

The ethical statistical practitioner:

1. Keeps informed about and adheres to applicable rules, approvals, and guidelines for the protection and welfare of human and animal subjects. Knows when work requires ethical review and oversight.[9]

2. Makes informed recommendations for sample size and statistical practice methodology in order to avoid the use of excessive or inadequate numbers of subjects and excessive risk to subjects

3. For animal studies, seeks to leverage statistical practice to reduce the number of animals used, refine experiments to increase the humane treatment of animals, and replace animal use where possible.

4. Protects people's privacy and the confidentiality of data concerning them, whether obtained from the individuals directly, other persons, or existing records. Knows and adheres to applicable rules, consents, and guidelines to protect private information.

5. Uses data only as permitted by data subjects' consent when applicable or considering their interests and welfare when consent is not required. This includes primary and secondary uses, use of repurposed data, sharing data, and linking data with additional data sets.

6. Considers the impact of statistical practice on society, groups, and individuals. Recognizes that statistical practice could adversely affect groups or the public perception of groups, including marginalized groups. Considers approaches to minimize negative impacts in applications or in framing results in reporting.

7. Refrains from collecting or using more data than is necessary. Uses confidential information only when permitted and only to the extent necessary. Seeks to minimize the risk of re-identification when sharing de-identified data or results where there is an expectation of confidentiality. Explains any impact of de-identification on accuracy of results.

8. To maximize contributions of data subjects, considers how best to use available data sources for exploration, training, testing, validation, or replication as needed for the application. The ethical statistical practitioner appropriately discloses how the data is used for these purposes and any limitations.

9. Knows the legal limitations on privacy and confidentiality assurances and does not over-promise or assume legal privacy and confidentiality protections where they may not apply.

---

[9] Examples of ethical review and oversight include an Institutional Review Board (IRB), an Institutional Animal Care and Use Committee (IACUC), or a compliance assessment.

10. Understands the provenance of the data, including origins, revisions, and any restrictions on usage, and fitness for use prior to conducting statistical practices.
11. Does not conduct statistical practice that could reasonably be interpreted by subjects as sanctioning a violation of their rights. Seeks to use statistical practices to promote the just and impartial treatment of all individuals.

Principle E: Responsibilities to members of multidisciplinary teams

Statistical practice is often conducted in teams made up of professionals with different professional standards. The statistical practitioner must know how to work ethically in this environment

The ethical statistical practitioner:

1. Recognizes and respects that other professions may have different ethical standards and obligations. Dissonance in ethics may still arise even if all members feel that they are working towards the same goal. It is essential to have a respectful exchange of views.
2. Prioritizes these Guidelines for the conduct of statistical practice in cases where ethical guidelines conflict.
3. Ensures that all communications regarding statistical practices are consistent with these Guidelines. Promotes transparency in all statistical practices.
4. Avoids compromising validity for expediency. Regardless of pressure on or within the team, does not use inappropriate statistical practices.

Principle F: Responsibilities to Fellow Statistical Practitioners and the Profession

Statistical practices occur in a wide range of contexts. Irrespective of job title and training, those who practice statistics have a responsibility to treat statistical practitioners, and the profession, with respect. Responsibilities to other practitioners and the profession include honest communication and engagement that can strengthen the work of others and the profession.

The ethical statistical practitioner:

1. Recognizes that statistical practitioners may have different expertise and experiences, which may lead to divergent judgments about statistical practices and results. Constructive discourse with mutual respect focuses on scientific principles and methodology and not personal attributes.
2. Helps strengthen, and does not undermine, the work of others through appropriate peer review or consultation. Provides feedback or advice that is impartial, constructive, and objective.
3. Takes full responsibility for their contributions as instructors, mentors, and supervisors of statistical practice by ensuring their best teaching and advising -- regardless of an academic or non-academic setting -- to ensure that developing practitioners are guided effectively as they learn and grow in their careers.
4. Promotes reproducibility and replication, whether results are "significant" or not, by sharing data, methods, and documentation to the extent possible.
5. Serves as an ambassador for statistical practice by promoting thoughtful choices about data acquisition, analytic procedures, and data structures among non-practitioners and students. Instills appreciation for the concepts and methods of statistical practice.

Principle G: Responsibilities of Leaders, Supervisors, and Mentors in Statistical Practice

Statistical practitioners leading, supervising, and/or mentoring people in statistical practice have specific obligations to follow and promote these Ethical Guidelines. Their support for – and insistence on – ethical statistical practice are essential for the integrity of the practice and profession of statistics as well as the practitioners themselves.

Those leading, supervising, or mentoring statistical practitioners are expected to:

1. Ensure appropriate statistical practice that is consistent with these Guidelines. Protect the statistical practitioners who comply with these Guidelines, and advocate for a culture that supports ethical statistical practice.
2. Promote a respectful, safe, and productive work environment. Encourage constructive engagement to improve statistical practice.

3. Identify and/or create opportunities for team members/mentees to develop professionally and maintain their proficiency.

4. Advocate for appropriate, timely, inclusion and participation of statistical practitioners as contributors/collaborators. Promote appropriate recognition of the contributions of statistical practitioners, including authorship if applicable.

5. Establish a culture that values validation of assumptions, and assessment of model/algorithm performance over time and across relevant subgroups, as needed. Communicate with relevant stakeholders regarding model or algorithm maintenance, failure, or actual or proposed modifications.

Principle H: Responsibilities Regarding Potential Misconduct

The ethical statistical practitioner understands that questions may arise concerning potential misconduct related to statistical, scientific, or professional practice. At times, a practitioner may accuse someone of misconduct, or be accused by others. At other times, a practitioner may be involved in the investigation of others' behavior. Allegations of misconduct may arise within different institutions with different standards and potentially different outcomes. The elements that follow relate specifically to allegations of statistical, scientific, and professional misconduct.

The ethical statistical practitioner:

1. Knows the definitions of, and procedures relating to, misconduct in their institutional setting. Seeks to clarify facts and intent before alleging misconduct by others. Recognizes that differences of opinion and honest error do not constitute unethical behavior.

2. Avoids condoning or appearing to condone statistical, scientific, or professional misconduct. Encourages other practitioners to avoid misconduct or the appearance of misconduct.

3. Does not make allegations that are poorly founded, or intended to intimidate. Recognizes such allegations as potential ethics violations.

4. Lodges complaints of misconduct discreetly and to the relevant institutional body. Does not act on allegations of misconduct without appropriate institutional referral, including those allegations originating from social media accounts or email listservs.

5. Insists upon a transparent and fair process to adjudicate claims of misconduct. Maintains confidentiality when participating in an

investigation. Discloses the investigation results honestly to appropriate parties and stakeholders once they are available.

6. Refuses to publicly question or discredit the reputation of a person based on a specific accusation of misconduct while due process continues to unfold.

7. Following an investigation of misconduct, supports the efforts of all parties involved to resume their careers in as normal a manner as possible, consistent with the outcome of the investigation.

8. Avoids, and acts to discourage, retaliation against or damage to the employability of those who responsibly call attention to possible misconduct.

Appendix

Responsibilities of organizations/institutions

Whenever organizations and institutions design the collection of, summarize, process, analyze, interpret, or present, data; or develop and/or deploy models or algorithms, they have responsibilities to use statistical practice in ways that are consistent with these Guidelines, as well as promote ethical statistical practice.

Organizations and institutions engage in, and promote, ethical statistical practice by:

1. Expecting and encouraging all employees and vendors who conduct statistical practice to adhere to these Guidelines. Promoting a workplace where the ethical practitioner may apply the Guidelines without being intimidated or coerced. Protecting statistical practitioners who comply with these Guidelines.

2. Engaging competent personnel to conduct statistical practice, and promote a productive work environment.

3. Promoting the professional development and maintenance of proficiency for employed statistical practitioners.

4. Supporting statistical practice that is objective and transparent. Not allowing organizational objectives or expectations to encourage unethical statistical practice by its employees.

5. Recognizing that the inclusion of statistical practitioners as authors, or acknowledgement of their contributions to projects or publications, requires their explicit permission because it may imply endorsement of the work.

6. Avoiding statistical practices that exploit vulnerable populations or create or perpetuate discrimination or unjust outcomes. Considering both scientific validity and impact on societal and human well-being that results from the organization's statistical practice.
7. Using professional qualifications and contributions as the basis for decisions regarding statistical practitioners' hiring, firing, promotion, work assignments, publications and presentations, candidacy for offices and awards, funding or approval of research, and other professional matters.

Those in leadership, supervisory, or managerial positions who oversee statistical practitioners promote ethical statistical practice by following Principle G and:

8. Recognizing that it is contrary to these Guidelines to report or follow only those results that conform to expectations without explicitly acknowledging competing findings and the basis for choices regarding which results to report, use, and/or cite.
9. Recognizing that the results of valid statistical studies cannot be guaranteed to conform to the expectations or desires of those commissioning the study or employing/supervising the statistical practitioner(s).
10. Objectively, accurately, and efficiently communicating a team's or practitioners' statistical work throughout the organization.
11. In cases where ethical issues are raised, representing them fairly within the organization's leadership team.
12. Managing resources and organizational strategy to direct teams of statistical practitioners along the most productive lines in light of the ethical standards contained in these Guidelines.

Some things to note, organized by Principle, are:

A. Professional Integrity and Accountability

Professional integrity and accountability require taking responsibility for one's work. Ethical statistical practice supports valid and prudent decision making with appropriate methodology. The ethical statistical practitioner represents their capabilities and activities honestly, and treats others with respect.

Note: this Principle (A) points out the need for respectful professional conduct as well as competence, judgment, diligence, self-respect, and how the ethical practitioner is worthy of the respect of other people.

B. Integrity of data and methods

The ethical statistical practitioner seeks to understand and mitigate known or suspected limitations, defects, or biases in the data or methods and communicates potential impacts on the interpretation, conclusions, recommendations, decisions, or other results of statistical practices.

Note: this Principle (B) addresses the need to ensure you possess, and then report, sufficient information to have (yourself) and give readers, including other practitioners, a clear understanding of the intent of the work, the provenance of the data, how and by whom any analysis was performed, and any limitations on the validity of results.

C. Responsibilities to Stakeholders

Those who fund, contribute to, use, or are affected by statistical practices are considered stakeholders. The ethical statistical practitioner respects the interests of stakeholders while practicing in compliance with these Guidelines.

Note: this Principle (C) discusses the practitioner's responsibility for assuring that statistical work is appropriate and also suitable to the needs and resources of those who are paying for it, that funders understand the capabilities and limitations of statistics in addressing their problem, and that the funder's confidential information is protected.

D. Responsibilities to research subjects, data subjects, or those directly affected by statistical practices, Data Subjects, or those directly affected by statistical practices

The ethical statistical practitioner does not misuse or condone the misuse of data. They protect and respect the rights and interests of human and animal subjects. These responsibilities extend to those who will be directly affected by statistical practices.

Note: This Principle (D) describes requirements for respectful professional conduct generally and also protecting the interests of human and animal subjects of research – as well as other uses of data. The responsibilities are important not only during data collection but also in the analysis, interpretation, and publication of the resulting findings.

E. Responsibilities to members of multidisciplinary teams

Statistical practice is often conducted in teams made up of professionals with different professional standards. The statistical practitioner must know how to work ethically in this environment.

Note: This Principle (E) addresses respectful professional conduct and the mutual responsibilities of professionals participating in multidisciplinary research teams.

F. Responsibilities to Fellow Statistical Practitioners and the Profession

Statistical practices occur in a wide range of contexts. Irrespective of job title and training, those who practice statistics have a responsibility to treat statistical practitioners, and the profession, with respect. Responsibilities to other practitioners and the profession include honest communication and engagement that can strengthen the work of others and the profession.

Note: Principle F focuses on the obligation to recognize the interdependence of professionals doing similar work, whether in the same or different organizations. Basically, ethical statistical practitioners must contribute to the strength of the profession by sharing nonproprietary data and methods, participating in peer review, and respecting differing professional opinions.

G. Responsibilities of Leaders, Supervisors, and Mentors in Statistical Practice

Statistical practitioners leading, supervising, and/or mentoring people in statistical practice have specific obligations to follow and promote these Ethical Guidelines. Their support for – and insistence on – ethical statistical practice are essential for the integrity of the practice and profession of statistics as well as the practitioners themselves.

Note: Principle G was newly integrated in the ASA Ethical Guidelines for 2022. It serves several purposes. Firstly, those who are studying statistics and data science -and these ethical guidelines – for the first time should recognize that, as they progress in their careers their responsibilities can also change. Secondly, as new practitioners consider internships, jobs, and new opportunities, they should be aware of the obligations that leaders, supervisors, and mentors have to provide an ethical and safe workplace. Finally, Principle G underlines the importance of ethical practice for the profession itself. Practitioners at all levels have responsibilities to practice ethically.

H. Responsibilities Regarding Potential Misconduct

The ethical statistical practitioner understands that questions may arise concerning potential misconduct related to statistical, scientific, or professional practice. At times, a practitioner may accuse someone of misconduct, or be accused by others. At other times, a practitioner may be involved in the investigation of others' behavior. Allegations of misconduct may arise within different institutions with different standards and potentially different outcomes. The elements that follow relate specifically to allegations of statistical, scientific, and professional misconduct.

Note: Principle H addresses the sometimes difficult process of investigating potential ethical or legal violations, and treating those involved, as well as the process, with both justice and respect.

Appendix. Responsibilities of Organizations/institutions

Whenever organizations and institutions design the collection of, summarize, process, analyze, interpret, or present, data; or develop and/or deploy models or algorithms, they have responsibilities to use statistical practice in ways that are consistent with these Guidelines, as well as promote ethical statistical practice.

Note: The Appendix is specific to *employers*, including organizations and institutions as well as leaders who are not themselves statistical practitioners. The Appendix encourages employers and clients to recognize the highly interdependent nature of statistical ethics and statistical validity. Employers and organizations must not pressure practitioners to produce a particular "result". They must avoid the potential social harm that can result from the dissemination of false or misleading statistical work. Organizations and institutions have an obligation to ensure that the context in which all are practicing supports ethical behavior.

We are still (in Section 1) focused on prerequisite knowledge, but to remind us that there are six KSAs (five others besides knowledge), we can take the opportunity to explore how the KSAs work together with the ASA Ethical Guideline Principles. Table 1.3.1 walks the reader through the six ER KSAs according to each of the eight main ASA Ethical Guideline Principles (rows).

Reading Table 1.3.1 across a single row outlines how, if ASA Guideline Principle A (professional integrity and accountability for example) is to be learned or applied, the analysis of any given example can be structured so that each of the steps in ethical reasoning are applied.

Reading Table 1.3.1 down a single column outlines how each of the ethical reasoning steps can be learned and practiced across cases or examples. For Guideline Principles D-F (Responsibilities to research subjects, data subjects, or those directly affected by statistical practices (D), research team colleagues (E) and to other statisticians/statistical practitioners (F)), the structures are similar enough that the description of the application of each ethical reasoning step is identical (collapsing across rows, not columns). However, each ethical reasoning step can and should be applied separately to considerations that are relevant to each specific obligation (to subjects; to team; to colleagues/other statisticians).

Ethical Reasoning Steps: / 2022 ASA Guideline Principle:	Identify/assess prerequisite knowledge	Identify relevant decision-making frameworks (e.g., virtue or utilitarianism)	Recognize an ethical issue (decision that must be made)	Identify and evaluate alternative actions	Make & justify a decision	Reflect on the decision
A. Professional integrity and accountability:	ASA GLs are professional practice standards that should be utilized in ethical reasoning. Other information could also be useful. (When in doubt, ask for help!). *	The way to decide how to resolve the ethical issue must involve a framework for weighing different options; **virtue** options (following the "ethical practitioner") and **utilitarianism**	Ethical issues in statistical practice arise whenever one or more ASA Guideline *Principles or their constituent elements* cannot be followed. Understanding "professional integrity and accountability"	Decisions that the ethical issue requires must also be identified – e.g., whether to ask for help, or do nothing. If decisions like, "follow the ASA GLs" or "fail to follow the ASA GLs" are under consideration, these alternatives can be evaluated using the	The default decision on ethical challenges can seem to be, "do nothing – and avoid that situation in the future". The ethical practitioner recognizes this cannot be justified by the	Reflecting on careful ethical decisions can maintain and promote the integrity of the profession. Understanding how the ethical challenge arose – is another way to reflect on the ethical decision making process,

Ethical Reasoning Steps: / 2022 ASA Guideline Principle:	Identify/assess prerequisite knowledge	Identify relevant decision-making frameworks (e.g., virtue or utilitarianism)	Recognize an ethical issue (decision that must be made)	Identify and evaluate alternative actions	Make & justify a decision	Reflect on the decision
A. Professional integrity and accountability continued....		(prioritizing the action that results in the least harm) are two straightforward methods.	is essential, because the ethical practitioner *always* has some responsibility to maintain the integrity of the profession.	decision-making frameworks. Virtue: follow the GLs! Utilitarian: harms are likely to accrue from failures to follow GLs, and any benefits will be minor.	ASA GLs, and often results in harms that may become worse because unacknowledged	and can further serve the profession or your colleagues (or both) by preventing future recurrences.

Ethical Reasoning Steps: 2022 ASA Guideline Principle:	Identify/ assess prerequisite knowledge	Identify relevant decision-making frameworks (e.g., virtue or utilitarianism)	Recognize an ethical issue (decision that must be made)	Identify and evaluate alternative actions	Make & justify a decision	Reflect on the decision
B. Integrity of data and methods	If the integrity or source of the data, or proscribed methods, cannot be ascertained, that constitutes an ethical challenge according to ASA GL B. Failures on ASA GL B can harm the	If a virtue framework highlights conflicting responsibilities to different stakeholders, utilitarianism can promote decisions that are consistent with the ASA GLs when the contributors of data are considered. *	Articulating (and then ensuring inclusion of) limitations and assumptions in reporting comprise decisions (about e.g., transparency in reporting) that can yield ethical challenges that are ethical for	The default alternatives can appear to be "acknowledge" and "do not acknowledge" limitations of the data and/or assumptions. Even acknowledging that this is a *decision* – and encouraging the ethical option- can support more ethical quantitative practice.	The justification for a decision to "not acknowledge" limitations may be "takes too much space", which can be effectively compared to the justification to acknowledge them: "this is ethical practice". Space constraints cannot take	Reflecting on the decision entails considering what went better and what could be improved for future engagement. The analyst can feel the least amount of control over data –but always retains control over the methods to be employed and how results

Ethical Reasoning Steps: 2022 ASA Guideline Principle:	Identify/ assess prerequisite knowledge	Identify relevant decision-making frameworks (e.g., virtue or utilitarianism)	Recognize an ethical issue (decision that must be made)	Identify and evaluate alternative actions	Make & justify a decision	Reflect on the decision
B. Integrity of data and methods continued...	public trust in science, as well as lead to bias and other harms.		the data analyst.		priority over ethical practice.	are presented. Utilizing the prerequisite knowledge can help inform colleagues.

Ethical Reasoning Steps: 2022 ASA Guideline Principle:	Identify/ assess prerequisite knowledge	Identify relevant decision-making frameworks (e.g., virtue or utilitarianism)	Recognize an ethical issue (decision that must be made)	Identify and evaluate alternative actions	Make & justify a decision	Reflect on the decision
C. Responsibilities to Stakeholders	Obligations to stakeholders can conflict themselves, leading to ethical challenges for the analyst. Prerequisite knowledge can include understanding all stakeholders and their perspectives on the data/analysis/results/interpretation.	Virtue ethics and utilitarianism function best when all stakeholders are correctly identified, and harms/benefits are fairly recognized; responsibilities can be prioritized *only* when all stakeholders are considered.	Ethical challenges can often arise when responsibilities to different stakeholders are in conflict. These can be compounded when career considerations are added.	Alternative actions must be concretely articulated to be evaluable, irrespective of the decision-making framework. Balancing and prioritizing responsibilities given the perspectives of all stakeholders can be challenging, so should be done systematically.	Making *and justifying* decisions is important when multiple stakeholder perspectives must be addressed. Justification - based on concrete evaluation of the perquisite knowledge- is crucial.	If an analyst chooses one stakeholder (e.g., science or the profession) to which the highest priority would be given most often, then reflection on decisions can follow naturally from the effects of making and justifying decisions that are consistent with this approach.

Ethical Reasoning Steps: / 2022 ASA Guideline Principle:	Identify/ assess prerequisite knowledge	Identify relevant decision-making frameworks (e.g., virtue or utilitarianism)	Recognize an ethical issue (decision that must be made)	Identify and evaluate alternative actions	Make & justify a decision	Reflect on the decision
D. Responsibilities to research subjects, data subjects, or those directly affected by statistical practices, Data Subjects, or those directly affected by statistical practices	Articulating prerequisite knowledge that is sufficient to identify and make informed and justifiable decisions may involve different sources of knowledge, but the assessment of	When balancing goods and harms (utilitarianism), consideration of research subjects, team colleagues, and other statistical practitioners can be helpful and informative. When using a virtue ethics	Ethical challenges can arise with respect to the treatment of research subjects (and colleagues in and outside the profession) but they are more likely to arise when responsibilities to these different	Although the perspectives of these three types of stakeholders may differ, the evaluation of alternative actions that are identified for a given ethical challenge can be synergistic. Decisions may be multifaceted or may proceed in stages; the extent to which the options are	The decision, as well as its justification, must balance the analyst's responsibilities to these three types of stakeholder.	Reflection on decisions that are made based on considerations of the analyst's responsibilities to stakeholders in these three groups may focus on the decision-making framework or how feasible it is to prioritize decision making in terms of each

Ethical Reasoning Steps:	Identify/ assess prerequisite knowledge	Identify relevant decision-making frameworks (e.g., virtue or utilitarianism)	Recognize an ethical issue (decision that must be made)	Identify and evaluate alternative actions	Make & justify a decision	Reflect on the decision
2022 ASA Guideline Principle:	it will be similar for all three of these Guideline Principles.	approach, it may be more helpful to choose one perspective to compare and contrast the effects of decisions on the other perspectives.	research participants come into conflict.	concretely articulated promotes thorough evaluation.		group (or all together).
E. Responsibilities to members of multidisciplinary teams						
F. Responsibilities to Fellow Statistical Practitioners and the Profession						

Ethical Reasoning Steps: / 2022 ASA Guideline Principle:	Identify/ assess prerequisite knowledge	Identify relevant decision-making frameworks (e.g., virtue or utilitarianism)	Recognize an ethical issue (decision that must be made)	Identify and evaluate alternative actions	Make & justify a decision	Reflect on the decision
G. Responsibilities of Leaders, Supervisors, and Mentors in Statistical Practice	Understanding the differences between the responsibilities of the practitioner and those of the leader, supervisor, or mentor is important for new practitioners. They should seek mentors who recognize their own	From the leader/mentor perspective, virtue ethics (what would the ideal practitioner do?) is a useful framework for them and for their team members. Utilitarianism can be useful for reasoning through statistical	It can be challenging to identify ethical issues that arise specifically from failures of others to recognize and act on their responsibilities as leaders. These GLs, together with organizational policies or codes of	The simplest alternative actions under this Guideline Principle are often "leader does" vs. "leader does not" follow their organization's ethical code. However, Principle G offers specific guidance that can be useful in cases where the organization's policy or code is less specific to	The justification for action that leverages this Principle could jointly depend on Principle F (Responsibili-ties to fellow practitioners and the profession).	Reflection on decisions relating to this Guideline Principle tend to be limited to focus on the program, institution, or context in which leaders' decisions were observed or experienced. However, this can still be useful and informative for the practitioner –if

Ethical Reasoning Steps: 2022 ASA Guideline Principle:	Identify/assess prerequisite knowledge	Identify relevant decision-making frameworks (e.g., virtue or utilitarianism)	Recognize an ethical issue (decision that must be made)	Identify and evaluate alternative actions	Make & justify a decision	Reflect on the decision
G. Responsibilities of Leaders, Supervisors, and Mentors in Statistical Practice continued...	obligations to practice ethically and to maintain a safe work environment where ethical practice is valued and modeled.	ethical challenges when the leader and the team jointly want to minimize harms.	conduct, can help to identify and articulate ethical challenges where this Guideline Principle pertains.	statistical practice. Additional alternatives could include "follow institutional policy" vs. "follow ASA Guidelines".		only for making future plans.

Ethical Reasoning Steps: 2022 ASA Guideline Principle:	Identify/ assess prerequisite knowledge	Identify relevant decision-making frameworks (e.g., virtue or utilitarianism)	Recognize an ethical issue (decision that must be made)	Identify and evaluate alternative actions	Make & justify a decision	Reflect on the decision
H. Responsibilities regarding potential misconduct	Prerequisite knowledge sufficient to engage this Guideline Principle competently and confidently may be one of the most crucial – for confident identification of real misconduct and confident	In addition to engaging the decision-making frameworks that include utilitarianism and virtue ethics, legal considerations may also come into play when determining a course of action relating to allegations of misconduct.	Ethical issues may arise from observation, allegation, or actual commission of misconduct, or activities that may be construed as such.	While the identification of alternative actions relating to these ethical issues may be clear cut (or proscribed), their evaluation can be very complex.	The justification of decisions about ethical issues arising from the observation, allegation, or actual commission of misconduct, or activities that may be construed as such must be explicit and must utilize evidence to the highest	Reflecting on decisions relating to the observation, allegation, or actual commission of misconduct, or activities that may be construed as such should emphasize both personal development (awareness of self and relation to the scientific or

Ethical Reasoning Steps:	Identify/ assess prerequisite knowledge	Identify relevant decision-making frameworks (e.g., virtue or utilitarianism)	Recognize an ethical issue (decision that must be made)	Identify and evaluate alternative actions	Make & justify a decision	Reflect on the decision
2022 ASA Guideline Principle:						
H. Responsibilities regarding potential misconduct continued...	identification of other behaviour that is not misconduct. *				degree of any of the Guideline Principles.	professional communities) *and* the program or process within which these responsibilities were engaged.

TABLE 1.3.1[10] : *Walking through the steps of Ethical Reasoning using the ASA Ethical Guidelines: Examples of executing ethical reasoning steps with each ASA (2022) Guideline Principle.*

Notes.

*Many institutions and businesses offer ethics/bioethics and even ombudsperson consultation to support and strengthen responsible conduct across scientific disciplines. These supports should be sought and utilized whenever ethical challenges arise that seem beyond the individual's ability to reason through or resolve.

The Appendix to the 2022 ASA Ethical Guidelines for Statistical Practice pertain to organizations and institutions. The appendix material is less directly linked to the individual practitioner's ethical reasoning, so those elements are omitted from this table.

The table is not comprehensive, although it may feel pretty massive. Whenever people say, or try to convince you, that "ethics is simply good citizenship", you might point to this table and note that there really is nothing simple about it!

---

[10] Table is adapted from Gunaratna NS & Tractenberg RE. (2016). Ethical reasoning with the 2016 revised ASA Ethical Guidelines for Statistical Practice. Proceedings of the 2016 Joint Statistical Meetings, Chicago, IL. Pp. 3763-3787. The contents are updated for 2022.

Chapter 1.4
The ACM Code of Ethics and Professional Conduct
Adopted by ACM Council 6/22/18.

Along a similar timeline followed by the ASA Ethical Guidelines for Statistical Practice, the Association for Computing Machinery (ACM) also formulated (1992) and revised (2018) a code of ethics (URL). Four main areas are elaborated on with specific behaviors that "a computing professional should" do. The Code is presented below, and is Copyrighted (c) 2018 by the Association for Computing Machinery.

Preamble

Computing professionals' actions change the world. To act responsibly, they should reflect upon the wider impacts of their work, consistently supporting the public good. The ACM Code of Ethics and Professional Conduct ("the Code") expresses the conscience of the profession.

The Code is designed to inspire and guide the ethical conduct of all computing professionals, including current and aspiring practitioners, instructors, students, influencers, and anyone who uses computing technology in an impactful way. Additionally, the Code serves as a basis for remediation when violations occur. The Code includes principles formulated as statements of responsibility, based on the understanding that the public good is always the primary consideration. Each principle is supplemented by guidelines, which provide explanations to assist computing professionals in understanding and applying the principle.

Section 1 outlines fundamental ethical principles that form the basis for the remainder of the Code. Section 2 addresses additional, more specific considerations of professional responsibility. Section 3 guides individuals who have a leadership role, whether in the workplace or in a volunteer professional capacity. Commitment to ethical conduct is required of every ACM member, and principles involving compliance with the Code are given in Section 4.

The Code as a whole is concerned with how fundamental ethical principles apply to a computing professional's conduct. The Code is not an algorithm for solving ethical problems; rather it serves as a basis for ethical decision-making. When thinking through a particular issue, a computing professional may find that multiple principles should be taken into account, and that different principles will have different relevance to the issue. Questions related to these kinds of issues can best be answered by thoughtful consideration of the fundamental ethical principles, understanding that the public good is the paramount consideration. The entire computing profession benefits when the ethical decision-making process is accountable to and transparent to all stakeholders. Open discussions about ethical issues promote this accountability and transparency.

1. General Ethical Principles

A computing professional should:

1.1 Contribute to society and to human well-being, acknowledging that all people are stakeholders in computing

This principle, which concerns the quality of life of all people, affirms an obligation of computing professionals, both individually and collectively, to use their skills for the benefit of society, its members, and the environment surrounding them. This obligation includes promoting fundamental human rights and protecting each individual's right to autonomy. An essential aim of computing professionals is to minimize negative consequences of computing, including threats to health, safety, personal security, and privacy. When the interests of multiple groups conflict, the needs of those less advantaged should be given increased attention and priority.

Computing professionals should consider whether the results of their efforts will respect diversity, will be used in socially responsible ways, will meet social needs, and will be broadly accessible. They are encouraged to actively contribute to society by engaging in pro bono or volunteer work that benefits the public good.

In addition to a safe social environment, human well-being requires a safe natural environment. Therefore, computing professionals should promote environmental sustainability both locally and globally.

1.2 Avoid harm

In this document, "harm" means negative consequences, especially when those consequences are significant and unjust. Examples of harm include unjustified physical or mental injury, unjustified destruction or disclosure of information, and unjustified damage to property, reputation, and the environment. This list is not exhaustive.

Well-intended actions, including those that accomplish assigned duties, may lead to harm. When that harm is unintended, those responsible are obliged to undo or mitigate the harm as much as possible. Avoiding harm begins with careful consideration of potential impacts on all those affected by decisions. When harm is an intentional part of the system, those responsible are obligated to ensure that the harm is ethically justified. In either case, ensure that all harm is minimized.

To minimize the possibility of indirectly or unintentionally harming others, computing professionals should follow generally accepted best practices unless there is a compelling ethical reason to do otherwise. Additionally, the consequences of data aggregation and emergent properties of systems should be carefully analyzed. Those involved with pervasive or infrastructure systems should also consider Principle 3.7.

A computing professional has an additional obligation to report any signs of system risks that might result in harm. If leaders do not act to curtail or mitigate such risks, it may be necessary to "blow the whistle" to reduce potential harm. However, capricious or misguided reporting of risks can itself be harmful. Before reporting risks, a computing professional should carefully assess relevant aspects of the situation.

1.3 Be honest and trustworthy

Honesty is an essential component of trustworthiness. A computing professional should be transparent and provide full disclosure of all pertinent system capabilities, limitations, and potential problems to the appropriate parties. Making deliberately false or misleading claims, fabricating or falsifying data, offering or accepting bribes, and other dishonest conduct are violations of the Code.

Computing professionals should be honest about their qualifications, and about any limitations in their competence to complete a task. Computing professionals should be forthright about any circumstances that might lead to either real or perceived conflicts of interest or otherwise tend to undermine the independence of their judgment. Furthermore, commitments should be honored.

Computing professionals should not misrepresent an organization's policies or procedures, and should not speak on behalf of an organization unless authorized to do so.

1.4 Be fair and take action not to discriminate

The values of equality, tolerance, respect for others, and justice govern this principle. Fairness requires that even careful decision processes provide some avenue for redress of grievances.

Computing professionals should foster fair participation of all people, including those of underrepresented groups. Prejudicial discrimination on the basis of age, color, disability, ethnicity, family status, gender identity, labor union membership, military status, nationality, race, religion or belief, sex, sexual orientation, or any other inappropriate factor is an explicit violation of the Code. Harassment, including sexual harassment, bullying, and other abuses of power and authority, is a form of discrimination that, amongst other harms, limits fair access to the virtual and physical spaces where such harassment takes place.

The use of information and technology may cause new, or enhance existing, inequities. Technologies and practices should be as inclusive and accessible as possible and computing professionals should take action to avoid creating systems or technologies that disenfranchise or oppress people. Failure to design for inclusiveness and accessibility may constitute unfair discrimination.

1.5 Respect the work required to produce new ideas, inventions, creative works, and computing artifacts

Developing new ideas, inventions, creative works, and computing artifacts creates value for society, and those who expend this effort should expect to gain value from their work. Computing professionals should therefore credit the creators of ideas, inventions, work, and artifacts, and respect copyrights, patents, trade secrets, license agreements, and other methods of protecting authors' works.

Both custom and the law recognize that some exceptions to a creator's control of a work are necessary for the public good. Computing professionals should not unduly oppose reasonable uses of their intellectual works. Efforts to help others by contributing time and energy to projects that help society illustrate a positive aspect of this principle. Such efforts include free and open-source software and work put into the public domain. Computing professionals should not claim private ownership of work that they or others have shared as public resources.

1.6 Respect privacy

The responsibility of respecting privacy applies to computing professionals in a particularly profound way. Technology enables the collection, monitoring, and exchange of personal information quickly, inexpensively, and often without the knowledge of the people affected. Therefore, a computing professional should become conversant in the various definitions and forms of privacy and should understand the rights and responsibilities associated with the collection and use of personal information.

Computing professionals should only use personal information for legitimate ends and without violating the rights of individuals and groups. This requires taking precautions to prevent re- identification of anonymized data or unauthorized data collection, ensuring the accuracy of data, understanding the provenance of the data, and protecting it from unauthorized access and accidental disclosure. Computing professionals should establish transparent policies and procedures that allow individuals to understand what data is being collected and how it is being used, to give informed consent for automatic data collection, and to review, obtain, correct inaccuracies in, and delete their personal data.

Only the minimum amount of personal information necessary should be collected in a system. The retention and disposal periods for that information should be clearly defined, enforced, and communicated to data subjects. Personal information gathered for a specific purpose should not be used for other purposes without the person's consent. Merged data collections can compromise privacy features present in the original collections. Therefore, computing professionals should take special care for privacy when merging data collections.

1.7 Honor confidentiality

Computing professionals are often entrusted with confidential information such as trade secrets, client data, nonpublic business strategies, financial information, research data, pre-publication scholarly articles, and patent applications. Computing professionals should protect confidentiality except in cases where it is evidence of the violation of law, of organizational regulations, or of the Code. In these cases, the nature or contents of that information should not be disclosed except to appropriate authorities. A computing professional should consider thoughtfully whether such disclosures are consistent with the Code.

2. Professional Responsibilities

A computing professional should:

2.1 Strive to achieve high quality in both the processes and products of professional work

Computing professionals should insist on and support high quality work from themselves and from colleagues. The dignity of employers, employees, colleagues, clients, users, and anyone else affected either directly or indirectly by the work should be respected throughout the process. Computing professionals should respect the right of those involved to transparent communication about the project. Professionals should be cognizant of any serious negative consequences affecting any stakeholder that may result from poor quality work and should resist inducements to neglect this responsibility.

2.2 Maintain high standards of professional competence, conduct, and ethical practice

High quality computing depends on individuals and teams who take personal and group responsibility for acquiring and maintaining professional competence. Professional competence starts with technical knowledge and with awareness of the social context in which their work may be deployed. Professional competence also requires skill in communication, in reflective analysis, and in recognizing and navigating ethical challenges. Upgrading skills should be an ongoing process and might include independent study, attending conferences or seminars, and other informal or formal education. Professional organizations and employers should encourage and facilitate these activities.

2.3 Know and respect existing rules pertaining to professional work

"Rules" here include local, regional, national, and international laws and regulations, as well as any policies and procedures of the organizations to which the professional belongs. Computing professionals must abide by these rules unless there is a compelling ethical justification to do otherwise. Rules that are judged unethical should be challenged. A rule may be unethical when it has an inadequate moral basis or causes recognizable harm. A computing professional should consider challenging the rule through existing channels before violating the rule. A computing professional who decides to violate a rule because it is unethical, or for any other reason, must consider potential consequences and accept responsibility for that action.

2.4 Accept and provide appropriate professional review

High quality professional work in computing depends on professional review at all stages. Whenever appropriate, computing professionals should seek and utilize peer and stakeholder review. Computing professionals should also provide constructive, critical reviews of others' work.

2.5 Give comprehensive and thorough evaluations of computer systems and their impacts, including analysis of possible risks

Computing professionals are in a position of trust, and therefore have a special responsibility to provide objective, credible evaluations and testimony to employers, employees, clients, users, and the public. Computing professionals should strive to be perceptive, thorough, and objective when evaluating, recommending, and presenting system descriptions and alternatives. Extraordinary care should be taken to identify and mitigate potential risks in machine learning systems. A system for which future risks cannot be reliably predicted requires frequent reassessment of risk as the system evolves in use, or it should not be deployed. Any issues that might result in major risk must be reported to appropriate parties.

2.6 Perform work only in areas of competence

A computing professional is responsible for evaluating potential work assignments. This includes evaluating the work's feasibility and advisability, and making a judgment about whether the work assignment is within the professional's areas of competence. If at any time before or during the work assignment the professional identifies a lack of a necessary expertise, they must disclose this to the employer or client. The client or employer may decide to pursue the assignment with the professional after additional time to acquire the necessary competencies, to pursue the assignment with someone else who has the required expertise, or to forgo the assignment. A computing professional's ethical judgment should be the final guide in deciding whether to work on the assignment.

2.7 Foster public awareness and understanding of computing, related technologies, and their consequences

As appropriate to the context and one's abilities, computing professionals should share technical knowledge with the public, foster awareness of computing, and encourage understanding of computing. These communications with the public should be clear, respectful, and welcoming. Important issues include the impacts of computer systems, their limitations, their vulnerabilities, and the opportunities that they present. Additionally, a computing professional should respectfully address inaccurate or misleading information related to computing.

2.8 Access computing and communication resources only when authorized or when compelled by the public good

Individuals and organizations have the right to restrict access to their systems and data so long as the restrictions are consistent with other principles in the Code. Consequently, computing professionals should not access another's computer system, software, or data without a reasonable belief that such an action would be authorized or a compelling belief that it is consistent with the public good. A system being publicly accessible is not sufficient grounds on its own to imply authorization. Under exceptional circumstances a computing professional may use unauthorized access to disrupt or inhibit the functioning of malicious systems; extraordinary precautions must be taken in these instances to avoid harm to others.

2.9 Design and implement systems that are robustly and usably secure

Breaches of computer security cause harm. Robust security should be a primary consideration when designing and implementing systems. Computing professionals should perform due diligence to ensure the system functions as intended, and take appropriate action to secure resources against accidental and intentional misuse, modification, and denial of service. As threats can arise and change after a system is deployed, computing professionals should integrate mitigation techniques and policies, such as monitoring, patching, and vulnerability reporting. Computing professionals should also take steps to ensure parties affected by data breaches are notified in a timely and clear manner, providing appropriate guidance and remediation.

To ensure the system achieves its intended purpose, security features should be designed to be as intuitive and easy to use as possible. Computing professionals should discourage security precautions that are too confusing, are situationally inappropriate, or otherwise inhibit legitimate use.

In cases where misuse or harm are predictable or unavoidable, the best option may be to not implement the system.

3. Professional Leadership Principles

Leadership may either be a formal designation or arise informally from influence over others. In this section, "leader" means any member of an organization or group who has influence, educational responsibilities, or managerial responsibilities. While these principles apply to all computing professionals, leaders bear a heightened responsibility to uphold and promote them, both within and through their organizations.

A computing professional, especially one acting as a leader, should:

3.1 Ensure that the public good is the central concern during all professional computing work

People—including users, customers, colleagues, and others affected directly or indirectly— should always be the central concern in computing. The public good should always be an explicit consideration when evaluating tasks associated with research, requirements analysis, design, implementation, testing, validation, deployment, maintenance, retirement, and disposal. Computing professionals should keep this focus no matter which methodologies or techniques they use in their practice.

3.2 Articulate, encourage acceptance of, and evaluate fulfillment of social responsibilities by members of the organization or group

Technical organizations and groups affect broader society, and their leaders should accept the associated responsibilities. Organizations—through procedures and attitudes oriented toward quality, transparency, and the welfare of society—reduce harm to the public and raise awareness of the influence of technology in our lives. Therefore, leaders should encourage full participation of computing professionals in meeting relevant social responsibilities and discourage tendencies to do otherwise.

3.3 Manage personnel and resources to enhance the quality of working life

Leaders should ensure that they enhance, not degrade, the quality of working life. Leaders should consider the personal and professional development, accessibility requirements, physical safety, psychological well-being, and human dignity of all workers. Appropriate human-computer ergonomic standards should be used in the workplace.

3.4 Articulate, apply, and support policies and processes that reflect the principles of the Code

Leaders should pursue clearly defined organizational policies that are consistent with the Code and effectively communicate them to relevant stakeholders. In addition, leaders should encourage and reward compliance with those policies, and take appropriate action when policies are violated. Designing or implementing processes that deliberately or negligently violate, or tend to enable the violation of, the Code's principles is ethically unacceptable.

3.5 Create opportunities for members of the organization or group to grow as professionals

Educational opportunities are essential for all organization and group members. Leaders should ensure that opportunities are available to computing professionals to help them improve their knowledge and skills in professionalism, in the practice of ethics, and in their technical specialties. These opportunities should include experiences that familiarize computing professionals with the consequences and limitations of particular types of systems. Computing professionals should be fully aware of the dangers of oversimplified approaches, the improbability of anticipating every possible operating condition, the inevitability of software errors, the interactions of systems and their contexts, and other issues related to the complexity of their profession—and thus be confident in taking on responsibilities for the work that they do.

3.6 Use care when modifying or retiring systems

Interface changes, the removal of features, and even software updates have an impact on the productivity of users and the quality of their work. Leaders should take care when changing or discontinuing support for system features on which people still depend. Leaders should thoroughly investigate viable alternatives to removing support for a legacy system. If these alternatives are unacceptably risky or impractical, the developer should assist stakeholders' graceful migration from the system to an alternative. Users should be notified of the risks of continued use of the unsupported system long before support ends. Computing professionals should assist system users in monitoring the operational viability of their computing systems, and help them understand that timely replacement of inappropriate or outdated features or entire systems may be needed.

3.7 Recognize and take special care of systems that become integrated into the infrastructure of society

Even the simplest computer systems have the potential to impact all aspects of society when integrated with everyday activities such as commerce, travel, government, healthcare, and education. When organizations and groups develop systems that become an important part of the infrastructure of society, their leaders have an added responsibility to be good stewards of these systems. Part of that stewardship requires establishing policies for fair system access, including for those who may have been excluded. That stewardship also requires that computing professionals monitor the level of integration of their systems into the infrastructure of society. As the level of adoption changes, the ethical responsibilities of the organization or group are likely to change as well. Continual monitoring of how society is using a system will allow the organization or group to remain consistent with their ethical obligations outlined in the Code. When appropriate standards of care do not exist, computing professionals have a duty to ensure they are developed.

4. Compliance with the Code

A computing professional should:

4.1 Uphold, promote, and respect the principles of the Code

The future of computing depends on both technical and ethical excellence. Computing professionals should adhere to the principles of the Code and contribute to improving them. Computing professionals who recognize breaches of the Code should take actions to resolve the ethical issues they recognize, including, when reasonable, expressing their concern to the person or persons thought to be violating the Code.

4.2 Treat violations of the Code as inconsistent with membership in the ACM

Each ACM member should encourage and support adherence by all computing professionals regardless of ACM membership. ACM members who recognize a breach of the Code should consider reporting the violation to the ACM, which may result in remedial action as specified in the ACM's Code of Ethics and Professional Conduct Enforcement Policy, which can be found here: https://ethics.acm.org/wp-content/uploads/2018/07/2018-ACM-Code-of-Ethics-Enforcement-Procedure.pdf

Referring to the 1992 version, Anderson et al. (1993) stated that "(c)ommitment to ethical professional conduct is expected of every voting, associate, and student member of ACM." In 2013, computer science curriculum guidelines (UG) were revised http://www.acm.org/education/CS2013-final-report.pdf and these new guidelines include a requirement for "core hours in the social issues and professional practice knowledge area" to help "to promote a greater understanding of the implications of social responsibility among students". And, "(c)urricula must prepare students for lifelong learning and must include professional practice (e.g., communication skills, teamwork, ethics) as components of the undergraduate experience." (p.21) Characteristics of graduates of these programs include "(c)ommitment to professional responsibility. Graduates should recognize the social, legal, ethical, and cultural issues inherent in the discipline of computing. They must further recognize that social, legal, and ethical standards vary internationally. They should be knowledgeable about the interplay of ethical issues, technical problems, and aesthetic values that play an important part in the development of computing systems. Practitioners must understand their individual and collective responsibility and the possible consequences of failure. They must understand their own limitations as well as the limitations of their tools." (p.25).

Just as we saw in the previous chapter on the ASA GLs, it can be helpful to walk through the ER KSAs with the ACM overall principles. This is done in Table 1.4.1.

Ethical Reasoning Steps: 2018 ACM Code of Ethics (CE) Principle:	Identify/ assess prerequisite knowledge	Identify relevant decision-making frameworks (e.g., virtue or utilitarianism)	Recognize an ethical issue (decision that must be made)	Identify and evaluate alternative actions	Make & justify a decision	Reflect on the decision
1. General Ethical Principles	To engage in ethical reasoning, ACM CE, plus stakeholder analysis (SHA), and possibly the ASA GLs, will be necessary (when in doubt, ask for help!). *	The framework for weighing different options and organizing prerequisite knowledge into evidence that helps recognize the ethical issue; *virtue* (guided by the CE) and **utilitarianism** are two straightforward frameworks.	Ethical issues in quantitative practice arise whenever one or more ACM CE Principles *or their constituent elements* cannot be followed. Also, if undue burden or harms accrue to stakeholders –or the risk of harms exist– this can create	Actions that the ethical issue requires may be identified by the ACM CE– e.g., whether to ask a colleague or mentor for help, or to share the ACM CE with the client/ collaborator, or to notify authorities that inappropriate use of data may occur. These alternatives can be evaluated using the	The obligation of computing professionals to avoid harm and "act to curtail or mitigate" risks means that "do nothing – and avoid that situation in the future" is never an alternative that the CE suggests, prerequisite knowledge plus the framework	The justification of a decision should be prepared in a way that can help alert others to the problem and a solution; understanding how the ethical challenge arose, and contemplating how to prevent recurrence, is another way to reflect on an

Ethical Reasoning Steps: 2018 ACM Code of Ethics (CE) Principle:	Identify/ assess prerequisite knowledge	Identify relevant decision-making frameworks (e.g., virtue or utilitarianism)	Recognize an ethical issue (decision that must be made)	Identify and evaluate alternative actions	Make & justify a decision	Reflect on the decision
1. General Ethical Principles continued…			bias/unfairness, which would be unethical.	decision-making frameworks.	can lead to both defensible decisions and their justification.	ethical decision. The purpose of reflection is to ensure the justification is sufficiently strong/well-reasoned, and to strengthen professional integrity for all practitioners –so that *all* can "avoid harm".

Ethical Reasoning Steps: 2018 ACM Code of Ethics (CE) Principle:	Identify/ assess prerequisite knowledge	Identify relevant decision-making frameworks (e.g., virtue or utilitarianism)	Recognize an ethical issue (decision that must be made)	Identify and evaluate alternative actions	Make & justify a decision	Reflect on the decision
2. Professional Responsibilities	ACM CE are professional practice standards that should be utilized in ethical reasoning. Other information (e.g., ASA GLs) could also be useful. (When in doubt, ask for help!). Professionals are obligated	The ACM Professional should follow the CE, in order to act in a professional manner (i.e., **virtue,** following the "ethical practitioner") and to avoid harm or the risk of harm (i.e., **utilitarianism,** prioritizing the action that	Ethical issues in quantitative practice arise whenever one or more ACM CE elements cannot be followed. Understanding "professional responsibilities" is essential, because the ethical practitioner *always* has some responsibility to maintain the	Decisions that the ethical issue requires must also be identified – e.g., whether to ask for help, "blow the whistle" or do nothing. If decisions like, "follow the ACM CE" or "fail to follow the ACM CE" are under consideration, these alternatives can be evaluated using the decision-making frameworks.	The default decision on ethical challenges can seem to be, "do nothing – and avoid that situation in the future". The ethical practitioner recognizes this cannot be justified by the ACM CE, and often results in harms that may become worse	Reflecting on careful ethical decisions can maintain and promote the integrity of the profession. Understanding how the ethical challenge arose given the context and possibly professional responsibilities of others on the team (if they are following a different CE) is

Ethical Reasoning Steps: / 2018 ACM Code of Ethics (CE) Principle:	Identify/ assess prerequisite knowledge	Identify relevant decision-making frameworks (e.g., virtue or utilitarianism)	Recognize an ethical issue (decision that must be made)	Identify and evaluate alternative actions	Make & justify a decision	Reflect on the decision
2. Professional Responsibilities continued…	to perform to the best of their abilities. *	results in the least harm).	integrity of the profession, and to do their job in an ethical manner so as to protect stakeholders from harm or the risk of harm.	Virtue: "follow the CE!" Utilitarian: "harms are likely to accrue from failures to follow CE, and any benefits will not be sufficient to justify harms or risks of harms."	because unacknowledged.	another way to reflect on the ethical decision-making process. "Not implementing a system" is a decision that requires reflection as well as justification, to serve the profession or your colleagues (or both) by preventing future recurrences.

Ethical Reasoning Steps: 2018 ACM Code of Ethics (CE) Principle:	Identify/assess prerequisite knowledge	Identify relevant decision-making frameworks (e.g., virtue or utilitarianism)	Recognize an ethical issue (decision that must be made)	Identify and evaluate alternative actions	Make & justify a decision	Reflect on the decision
3. Professional Leadership Principles	Leaders in computing are responsible for ensuring their organization and/or team understand their obligations to follow the ACM CE. Other information could be useful in helping team members understand and act in accordance	The public good, and avoiding harms, suggest that the ACM Professional should utilize the SHA to outline harms and ultimately, take a **utilitarianism** perspective in decision making, so that decisions will result in the least harm.	Ethical issues may arise for computing leaders when organizational (or work team) policies conflict with ACM CE, or otherwise prevent other computing professionals to fail to comply with the ACM CE. The ethical practitioner	The leader in particular has an obligation to empower all team members to reason ethically at work, so they can apply (or learn to apply) the ACM CE in the workplace. Recognizing that "do nothing" – particularly if harms or potential harms are recognized – is unethical but alternatives to this	The ethical practitioner recognizes that harms and risks of harms cannot be tolerated, so "do not look for harms", "do not raise alarms" and other non-actions are never justified by the ACM CE. Leaders can support brainstorming about harms and walk team	Reflecting on ethical decisions can maintain and promote the integrity of the profession, and leaders can encourage this by modeling reflection and ethical reasoning.

Ethical Reasoning Steps: 2018 ACM Code of Ethics (CE) Principle:	Identify/assess prerequisite knowledge	Identify relevant decision-making frameworks (e.g., virtue or utilitarianism)	Recognize an ethical issue (decision that must be made)	Identify and evaluate alternative actions	Make & justify a decision	Reflect on the decision
3. Professional Leadership Principles continued...	with the CE, particularly in terms of identifying harms/potential harms and acting to mitigate those. The leader knows that the ACM CE advo-cates whistle-blowing and abandoning systems when their misuse or harms are predictable or unavoidable.*		*always* has some responsibility to maintain the integrity of the profession, and to do their job in an ethical manner so as to protect stakeholders from harm or the risk of harm; so the leader is also responsible for ensuring/ empowering all team members to do so.	action (inaction) can be challenging to formulate. Conversations can support identification and evaluation of plausible alternatives (which may include, "do not implement the system").	members through the justifications for decisions to reinforce the ethical reasoning of all practitioners.	

Ethical Reasoning Steps: / 2018 ACM Code of Ethics (CE) Principle:	Identify/ assess prerequisite knowledge	Identify relevant decision-making frameworks (e.g., virtue or utilitarianism)	Recognize an ethical issue (decision that must be made)	Identify and evaluate alternative actions	Make & justify a decision	Reflect on the decision
4. Compliance with the Code	The ACM CE pertains to all who use computing. "Computing professionals who recognize breaches of the Code should take actions to resolve the ethical issues they recognize, including, when reasonable,	The public good, and "the future of computing", depend on ethical excellence, suggesting that the ACM Professional should take a **virtue** perspective, i.e., "each ACM member should encourage and	When computing professionals fail to follow the CE, it creates an ethical issue for all computing professionals who observe it. The ethical practitioner *always* has some responsibility	The CE specifies the alternative actions on this principle: "take actions to resolve the ethical issues they recognize"; and/or, "when reasonable, expressing their concern to the person or persons thought to be violating the Code"; and/or "consider reporting the violation to the	While the ACM CE specifies the alternatives actions for professionals who recognize breaches of the CE, it also recognizes that not all practitioners will be in a position to express their concern to the person/s violating the CE;	Understanding how compliance with the CE represents commitment to the profession as well as the public good is an important aspect of the professional identity of the computing practitioner. Reflection on decisions about

Ethical Reasoning Steps: / 2018 ACM Code of Ethics (CE) Principle:	Identify/ assess prerequisite knowledge	Identify relevant decision-making frameworks (e.g., virtue or utilitarianism)	Recognize an ethical issue (decision that must be made)	Identify and evaluate alternative actions	Make & justify a decision	Reflect on the decision
4. Compliance with the Code continued…	expressing their concern to the person or persons thought to be violating the Code."	support adherence by all computing professionals regardless of ACM membership". With more experience with the CE, utilitarianism and other decision-making perspectives may become more viable.	to maintain the integrity of the profession, so failing to take actions to resolve such a violation, if they are feasible, is itself a violation of the CE.	ACM." "Take no action" is never a plausible alternative.	they may also not be confident enough to report the violation to the ACM. Justification of the decision of which of these alternatives was chosen may be reflected on for future action.	Principle 4 may change as the practitioner advances in their career – and particularly as the practitioner transitions into positions of leadership.

Table 1.4.1.: *Walking through the steps of Ethical Reasoning using the ACM Code of Ethics: Examples of executing ethical reasoning steps with each ACM (2018) Code of Ethics Principle.*

Similar to Table 1.3.1 walking through the 2022 ASA GLs, Table 1.4.1 showing examples of ethical reasoning with the ACM CE is not comprehensive, although it may seem complex. It is impossible to consider every possible time you would need to reason ethically, so the point here (and with Table 1.3.1) is to point out that every point of the framework (ASA GLs or ACM CE) can be reasoned with or through. You can see that, while the ACM CE is shorter than the ASA GLs, the steps of ethical reasoning can still be applied and yield a rich set of considerations. The perspective of this book is that it is worthwhile to take the time to orient yourself to both reasoning and the practice standards, so that as you progress in your career and your responsibilities and sophistication in practice grow, so will your ability to both reason *and practice* ethically.

Chapter 1.5
Prerequisite knowledge that is common to both ACM CE and ASA GLs

As you consider the ethical practice standards presented in the previous two chapters, their similarities and differences bear some consideration. For example, the ASA does not have an enforcement policy, while the ACM does. Note also that the ACM defines "harm": *In this document, "harm" means negative consequences to any stakeholder, especially when those consequences are significant and unjust.*" The ASA does not include a definition like this, nor does the ASA stipulate that an ethical statistician is "honest" or "trustworthy" – primarily because these are not observable, not because they are not desirable. Finally, note that both sets of standards describe what the ethical practitioner DOES or should DO, representing a virtue perspective; however, the ACM CE stipulates repeatedly that the public good should be prioritized, and that harms must be minimized, which is a fairly explicit utilitarian perspective for ethical decision making. It *is* possible to follow the ACM "norms" of ethical behavior, i.e., use a virtue approach, whereby harms are minimized. The two perspectives are not incompatible! It is simply pointed out that the ASA is describing how the ethical practitioner behaves, or can be recognized; the ACM is describing the professional's obligation to minimize harm whenever possible, even if that means that a system should not be developed or deployed.

The most recent versions of both of these professional guidelines were presented in full in Chapters 1.3 (ASA Ethical Guidelines for Statistical Practice (GLs, 2022)) and 1.4 (ACM Code of Ethics (CE, 2018)). As you've seen, they are comprehensive, but you also saw that the principles and constituent elements are organized differently. It can help us to understand both the features of professional practice that the ASA and ACM prioritize as essential to ethical behavior, and have a stronger idea of exactly what it means to be a "statistician", a "data scientist", or a "statistician/data scientist" if we look at the features of these two practice standards together. The domain of data science is dependent on statistics and computing, and as both the American Statistical Association (ASA), representing roughly 18,000 practitioners worldwide. The Association for Computing Machinery (ACM), representing roughly 100,000 computing professionals worldwide, assert that their ethical practice guidance should pertain to members and non-members alike who utilize their methods and techniques – promoting the ethical use of statistical and computing practices; thus, these are a natural first step for understanding and describing **ethical data science**.

ASA Ethical Guidelines

As you saw in Chapter 1.3, the ASA Ethical Guidelines for Statistical Practice (ASA, 2022) includes 60 items under eight general areas, plus a 12-item Appendix:

A. Professional Integrity & Accountability (12)
B. Integrity of data and methods (7)
C. Responsibilities to Stakeholders (8)
D. Responsibilities to research subjects, data subjects, or those directly affected by statistical practices, Data Subjects, or those directly affected by statistical practices (11)
E. Responsibilities to members of multidisciplinary teams (4)
F. Responsibilities to Fellow Statistical Practitioners and the Profession (5)
G. Responsibilities of Leaders, Supervisors, and Mentors in Statistical Practice (5)
H. Responsibilities regarding potential misconduct (8)
APPENDIX: Responsibilities of organizations/institutions (12)

Two things are noteworthy in the list above: firstly, that there are 12 different elements in Principle A "Professional Integrity & Accountability" – more than any other principle. The next most numerous Principle is D, "Responsibilities to research subjects, data subjects, or those directly affected by statistical practices, Data Subjects, or those directly affected by statistical practices" with 11 elements. These features reflect the fact that ethical practice is as dependent on professionalism and accountability (A) as it is on the ethical treatment of those whose data are accessed, analyzed, and utilized (D). The second thing to notice in the summary list of the ASA GLs is that there is a whole principle (G) focusing on ethical leadership/supervision/mentorship of statistical practice, along with an Appendix that is entirely dedicated to what employers (individuals, organizations, and institutions) of anyone who does/uses statistics should know, understand, and expect. Tractenberg (2020) reports on the near-complete alignment between the *2018* version of the ASA Ethical Guidelines and those for the Royal Statistical Society (RSS, 2014) and International Statistics Institute (ISI, 2010). The alignment is even tighter with the 2022 version of the Guidelines.

It is important to note that, while both the RSS and ISI have invested significant effort in their respective professional ethical statements, neither has articulated that their ethical guidelines pertain to non-members or to all those who utilize statistical methods and/or techniques. By contrast, the ASA Guidelines explicitly articulate that, "the term "statistical practitioner" includes all those who engage in statistical practice, regardless of job title, profession, level, or field of degree." The Guidelines promote ethical science and ethical evidence-driven decisions, and are intended "to help statistical practitioners make decisions ethically". The ASA Ethical Guidelines for Statistical Practice represent a "virtue ethics" perspective: ethical statistical practice is described as "what the ethical statistical practitioner would do" or how the ethical practitioner does his/her work. The ASA Guidelines represent a formal structure to promote the conduct, aims, and qualities that characterize or mark ethical practitioners who use statistics (but are not statisticians themselves), as well as ethical professional statisticians. With 72 specific elements, the ASA Guidelines are also more extensive than those of the other statistical organizations. Because of the near-total alignment between the ASA Guidelines and those of the RSS and ISI, the ASA Guidelines are the ones featured in the alignment explorations with the ACM Code that follow.

The ASA Guidelines were revised (2016) to support professionalism, which is defined by Merriam Webster as, "…the skill, good judgment, and polite behavior that is expected from a person who is trained to do a job well." ASA members and nonmembers alike are obliged to promote understanding, through trust in their professionalism and judgment, when they engage in data analysis. Specifically, the ASA Guidelines articulate, in the Preamble, that they exist "(t)o inform those relying on statistical analysis, including employers, colleagues and the public, of the standards that they should expect."

The ASA Ethical Guidelines for Statistical Practice are intended to support every practitioner in the quantitative sciences –whether trained as a statistician or as a scientist or other professional with quantitative expertise. The 2016 revision of the Guidelines (maintained in the 2018 and 2022 revisions) represent the habits of mind that characterize the experienced – ethical - quantitative practitioner and as such, they were specifically intended to be disseminated and used in teaching and training, and can promote professionalism, and/or the development of professional identity, in all who use statistics.

ACM Code of Ethics

The ACM Code of Ethics was first published in 1992 and most recently revised in 2018 (ACM, 2018). The ACM Code represents a "utilitarian ethics" perspective: it directs "computing professionals" to focus attention on the positive and negative effects of their decisions – and emphasizes decisions that avoid or minimize harms. In the preamble, it is stated that computing professionals act responsibly when "consistently supporting the public good". However, because the ACM Code also specifies what "a computing professional *should*" do (in order to comply with the Code), it can also be interpreted from a virtue perspective, i.e., by describing the ideal professional (an ethical one). There are four main areas with specific points elaborated in terms of what "a computing professional should" do:

1. General ethical principles (7)
2. Professional responsibilities (9)
3. Professional leadership principles (7)
4. Compliance with the code (2)

Like the ASA Guidelines, the ACM Code seeks to support ethical decision making rather than to establish punishment, or describe explicitly what is right or wrong about aspects of practice. In the preamble, it is noted that,

> "The Code as a whole is concerned with how fundamental ethical principles apply to a computing professional's conduct. The Code is not an algorithm for solving ethical problems; rather it serves as a basis for ethical decision-making. When thinking through a particular issue, a computing professional may find that multiple principles should be taken into account, and that different principles will have different relevance to the issue."

Additionally, the Code states that (it is) "designed to inspire and guide the ethical conduct of all computing professionals, including current and aspiring practitioners, instructors, students, influencers, and anyone who uses computing technology in an impactful way." Following the 2018 revision of the ACM Code, a book was compiled to support teaching computing professionals about the Code; this is freely available on the ACM website (https://www.acm.org/binaries/content/assets/about/acm-code-of-ethics-booklet.pdf; individual case studies are also available, https://www.acm.org/code-of-ethics/case-studies). The ACM Code of Ethics is included (in full) in Chapter 1.4. Also, in Table 1.5.1, each of the 25 ACM elements is examined for thematic alignment with the eight principles of the ASA Guidelines.

As they are representative of two essential constituent disciplines for data science, Table 1 explores the thematic alignment of the ACM Code (rows) and the ASA Guideline Principles (columns). ACM Principle elements are given, and alignment with the ASA Principle is indicated in the cell, with the specific ASA element ("A" means concordance is with the A Principle statement; "A1" means concordance is exact or directly comparable with the first element under Principle A, etc.). The specific ASA element or Principle is indicated if there is a thematic or exact match to the specific element of the ACM Code. Although organizations and institutions (employers) are discussed in the Appendix, they are grouped together with Principle G because of the potential overlap of employers or leaders who are not statistical practitioners (Appendix) and leaders, supervisors, and mentors who are statistical practitioners (Principle G).

If the alignment is abstract (e.g., if the core idea is "professional integrity" but this is demonstrated substantially differently in statistics -e.g., with alternative sampling and analysis plans; and in computing machinery -e.g., with explicit evaluation of risks), then the alignment with the particular ASA element is indicated by parenthesis "(*element*)". That is, *abstract* concordance means the ACM Principle is described in terms of computing specifically, and would be demonstrated by/is described fundamentally differently by the ASA Guidelines. If there is neither exact nor thematic alignment, then the cell is left blank.

RESPONSIBILITIES TO/REGARDING:

ASA:	A Professional Integrity and Accountability	B Integrity of data and methods	C Stake-holders	D Research Subjects/Data Subjects and those affected by statistical practices	E Inter-disciplinary Team Members	F Other Practitioners/ Profession	G Leader/ supervisor/ mentor and APPENDIX	H Allegations of Potential Misconduct
ACM:								
ACM 1. GENERAL MORAL PRINCIPLES. *A computing professional should…*								
1.1 Contribute to society and to human well-being, acknowledging that all people are stakeholders in computing.	A2 and preamble	B6	C; C2, C6	D; D2, D6, D8, D11	E4	F4	G1, G5 Appendix 6	

ACM 1. GENERAL MORAL PRINCIPLES.
A computing professional should…

	A	B	C	ALL D	E	F	G	H
1.2 Avoid harm. *In this document, "harm" means negative consequences to any stakeholder, especially when those consequences are significant and unjust.*	A; A1-A4, A7-A9	B1-B3, B5, B6	C; C1-C5, C7-C8	ALL D	E1, E3, E4	F2-F4	G1, G2, G5 Appendix 6	H2, H3, H8
1.3 Be honest and trustworthy.	A; A1-A8, A11-A12.	B1-B5	C; C1, C2, C3	ALL D	E2, E3, E4	F4	G1, G2, G5 Appendix 1,2, 4-12	ALL H
1.4 Be fair and take action not to discriminate.	A2-A4, A5-A6, A8, A9, A12	B1-B4, B6-B7	C4, C7	D5-D6, D10-D11	E1	F1-F3	G1- G4 Appendix 1, 7,10	ALL H

ACM 1. GENERAL MORAL PRINCIPLES.
A computing professional should…

	A5	B4	C7		F2	G4 Appendix 5, 10	
1.5 Respect the work required to produce new ideas, inventions, creative works, and computing artifacts.							
1.6 Respect privacy.		B4	C7	D4, D5, D7, D9-D11			
1.7 Honor confidentiality.		B4	C7	D4, D5, D7, D9-D11			H4-H6

2. PROFESSIONAL RESPONSIBILITIES.

A computing professional should…

	A	B	C	D	E	F	G	H
2.1 Strive to achieve high quality in both the process and products of professional work.	ALL A	ALL B	ALL C	D10	E2, E3, E4	ALL F	ALL G; Appendix 1, 2, 4, 8-10, 12	
2.2 Maintain high standards of professional competence, conduct, and ethical practice.	ALL A	ALL B	ALL C	ALL D	E2, E3, E4	ALL F	ALL G; ALL Appendix	ALL H
2.3 Know, respect, and apply existing rules pertaining to professional work.	A1, A4-A8, A11-A12	B1, B4, B5	C2, C4-C8	D1-D5, D7, D9-D11	E2-E4	F4, F5	G1, G2; Appendix 1, 4, 7, 8-11	ALL H
2.4 Accept and provide appropriate professional review.	A1-A4	B5, B6		D1, D11	E3	F2	G2, G5; Appendix 3, 7	H2

2. PROFESSIONAL RESPONSIBILITIES.

A computing professional should…

	(A1, A4)	(B1, B2, B6)	(C4, C5)	(D6, D9-D11)	(E4)		(G5)	(H1)
2.5 Give comprehensive and thorough evaluations of computer systems and their impacts, including analysis of possible risks.							(Appendix 6)	
2.6 Have the necessary expertise, or the ability to obtain that expertise, for completing a work assignment before accepting it. Once accepted, that commitment should be honored.	A1, A2, A5, A10	B2	C1, C3	D1 (D9, D10)		F3	Appendix 1, 2	

2. PROFESSIONAL RESPONSIBILITIES.
A computing professional should…

2.7 Improve public awareness and understanding of computing, related technologies, and their consequences.	(A12)	(B1-B3, B5)	(C6)		(E3)	(F4, F5)	(G1, G5) (Appendix 9)	
2.8 Access computing and communication resources only when authorized to do so.		(B4)	(C7)	(D4, D5, D7, D11)	(E4)			(H5, H6)
2.9 Design and implement systems that are robustly and usably secure.	(A2, A4)	(B6)	(C2, C7)	(D4, D9, D10)	(E4)		(G5) (Appendix 1, 4)	(H5)

3. PROFESSIONAL LEADERSHIP PRINCIPLES.

In this section, "leader" means any member of an organization or group who has influence, educational responsibilities, or managerial responsibilities. These principles generally apply to organizations and groups, as well as their leaders.
A computing professional should...

3.1 Ensure that the public good is the central concern during all professional computing work.	(A2-A4, A8)	(B1, B4, B5)	(C2, C6)	(D1, D2, D6; D10, D11)	(E3, E4)	(F2, F3)	(G1, G2, G5) Appendix 4, 6	(H2, H3)
3.2 Articulate, encourage acceptance of, and evaluate fulfillment of the social responsibilities of members of an organization or group.	(A9)		(C6, C8)	(D11)		(F3)	(G2, G5) Appendix 4, 6	
3.3 Manage personnel and resources to enhance the quality of working life.	(A4, A8, A9, A12)	(B6)	(C2)			F3	ALL G Appendix 3, 10, 12	ALL H

3. PROFESSIONAL LEADERSHIP PRINCIPLES.

In this section, "leader" means any member of an organization or group who has influence, educational responsibilities, or managerial responsibilities. These principles generally apply to organizations and groups, as well as their leaders.
A computing professional should....

	(A4, A8, A9, A12)	(ALL B)	(C2, C4, C5, C8)	(ALL D)	(E2 – E4)	(F2-F5)	(ALL G)	(ALL H)
3.4 Articulate, apply, and support policies and processes that reflect the principles in the Code.							Appendix 1, 4, 6, 8-12	
3.5 Create opportunities for members of the organization or group to learn and be accountable for the scope, functions, limitations, and impacts of systems.	(A1)	(B6)	(C1)			(F2, F5)	G4 Appendix 3	
3.6 Retire legacy systems with care.		(B6)						
3.7 Recognize when a computer system is becoming integrated into the infrastructure of society, and adopt an appropriate standard of care for that system and its users.								

4. COMPLIANCE WITH THE CODE. *A computing professional should…*	(* suggested in the ASA preamble; included *by implication* only (that "the ethical statistical practitioner…" follows the GLs))			
4.1 Uphold, promote, and respect the principles of the Code.	A9, A12		E2, E3		G1
4.2 Treat violations of the Code as inconsistent with membership in the ACM.					Appendix 1, 8, 12

Table 1.5.1. *Association of Computing Machinery Ethics (2018) Code v3 vs American Statistical Association Ethical Guidelines (2022)*

Table 1.5.1 shows a high level of *thematic concordance* between the two practice standards. That is, while the wording (and some of the manifestations of the specific feature of ethical practice) is different, both organizations specify similar features as foundational in ethical practice with the methods and materials of their respective disciplines. Several ACM Code elements are aligned with **every** ASA principle; these are highlighted in light grey.

Of the 25 ACM Code elements, two of them are not aligned with any aspect of the ASA Guidelines (highlighted in dark grey in Table 1.5.1). These are unaligned for good reasons: in the first case, ACM item 3.7, "computer systems" can easily become integrated into social or business infrastructure – sometimes to the point that only the computing professionals with extensive experience with the systems know exactly how they work. This is rarely the case for statistical models, although that is becoming less clear cut as algorithms are developed to estimate probabilities associated with outcomes based on ever-changing data sets, across businesses worldwide. However, "the system and its users", might never be as clear for algorithms as they are for computing systems. Thus, the lack of alignment with specific ASA elements is understandable. The second case is far more concrete: item 4.2 implicitly refers to a process for identifying and adjudicating violations of the ACM Code of Ethics, and the ASA has no such process. Neither the ASA Committee on Professional Ethics, nor any other ASA body, has no responsibilities or power to rule on potential violations of the ASA Ethical Guidelines. Thus, there is no alignment between the ASA and ACM on this item. It is entirely up to the statistical practitioner to understand and follow the ASA Ethical Guidelines for Statistical Practice.

Beyond these two elements, every ASA Guideline has at least some thematic alignment with all of the other ACM elements. While ACM Principle 4, "compliance with the code" has some matches in the ASA GLs, with multiple specific mentions encouraging compliance with the Guidelines (in Principles A, E, G, and the Appendix), as well as several mentions of not letting others or their discipline-specific guidelines or norms conflict with the ASA GLs, the ACM Code goes further, specifically stating that violations of their Code should be considered "inconsistent with membership in the ACM". Since the Guidelines are intended for all "who use statistical methods in their professional work" and not just statistical practitioners who are ASA members, it may not matter to some practitioners whether failing to follow the ASA Guidelines is inconsistent with *membership* in the ASA. As of the 2022 revisions of the ASA Ethical Guidelines, a new item appears in Principle A, (the ethical statistical practitioner) "Upholds, promotes, and respects these Guidelines." (A12). Additionally, the ASA GL Preamble states that, "To justify unethical behaviors, or to exploit gaps in the Guidelines, is unprofessional, and inconsistent with these Guidelines."

We have seen that there is a great deal of overlap in terms of these professional cultural norms, particularly for the ASA and ACM – including their respective statements that the standards for ethical practice apply to both members and to those who utilize the tools and techniques belonging to their respective domains. Both organizations articulate the need to be professional and respectful in practice, whether users of statistics or computing are members or not, or hold specific job titles ("statistician", "computing professional"), or have different job titles but use these domains in their work.

While the need to be respectful, and ethical in the use of the tools and techniques of each discipline or profession have not changed over time, the professional practice standards are dynamic, requiring review and revision periodically to ensure that they remain both relevant for how professionals practice (statistics/data science/computing) and also for what the professional cultural norms for ethical behavior are. Interested readers who are members of the ASA are encouraged to contribute to the ongoing effort to ensure that the ASA Ethical Guidelines reflect *current* practice considerations, by carefully reading the GLs and suggesting any revisions they feel should be contemplated for the next 5-yearly update. Input from ASA members in good standing as to updating the ASA Ethical Guidelines are welcomed at https://community. amstat.org/communities/community-home?CommunityKey=b482848a-43f9-441c-84cb-96bfe4f732c7.

The ASA as an organization does not intend for membership to be contingent on following the ASA GLs, so it is not feasible for the ASA to describe failures to follow the ASA GLs as "inconsistent with ASA membership". Not all statistical societies have this perspective; the Royal Statistical Society in the UK (http://www.rss.org.uk/Images/PDF/join-us/RSS-Code-of-Conduct-2014.pdf, Royal Statistical Society, 2014) *does* suggest this to be the case ("The code is...recommended to all fellows of the Society"), similar to the ACM. Instead, the ASA GLs' Preamble (purpose of the GLs) suggest that the *ethical practitioner* follows the ASA GLs, and throughout the GLs it is implied that, if the statistician does *not* follow each ASA GL principle, then they are not considered "the ethical statistical practitioner...". Moreover, the ASA GL Preamble states, "The principles expressed here should guide both those whose primary occupation is statistics and those in all other disciplines who use statistical methods in their professional work." That is, the ASA GLs are not limited to ASA members or professional statisticians, but like the ACM, the GLs are intended for all practitioners of statistics (ACM 4.2, "Each ACM member should encourage and support adherence by all computing professionals *regardless of ACM membership*", emphasis added). The Royal Statistical Society is more targeted, stating in their 2014 revision that their "code of conduct" "defines the actions and behavior expected of RSS fellows practicing in their everyday professional life and has been drawn up to reflect the standards of conduct and work expected of *all practicing statisticians* (emphasis added)."

It is important to notice that there are some ACM elements that have what might be called "conditional" alignment, denoted with parentheses (e.g., "(E.2)") in the intersecting cell, meaning that the two elements are aligned, but only under specific circumstances. These are more fully explored in Tractenberg (2020). Given the concordance of these two sets of practice standards from 2018 and 2022, they represent an excellent – coherent – resource for beginning to describe what constitutes ethical data science. There is, perhaps, a surprising amount of overlap given that these two sets of practice guidelines originated in the 1990s, when practitioners in statistics and computing machinery had a bit less in common than they do in 2022. The strongest overlap with the ASA (with yellow indicating total agreement on all ASA Principles) is for the "general ethical principles (ACM Principle 1).

Therefore, when we consider *what constitutes prerequisite knowledge*, there are certain types and sources of "knowledge" that are prerequisites to embarking on an analysis or assessment of what ethical actions are called for, what decisions had been or need to be made, and how to make those decisions as ethically as possible. Whenever you are faced with making a decision about how to practice ethically, or what to do in response to behavior, decisions, or situations where you might wonder (or be convinced) whether or not they are ethical, assembling all the available information – your prerequisite knowledge about what to do/do next – is essential. The ASA and ACM practice standards are viable, dynamic, and *normative* sources of knowledge that you can always bring to bear to determine what to do in order to make an ethical decision, or to determine whether or not some action/decision or behavior are ethical.

The ASA Ethical Guidelines for Statistical Practice and ACM Code of Ethics both represent a "virtue ethics" decision making framework, which can generally be summarized as, "what would the ethical practitioner do in this case?" Either the ASA or ACM standards can be used to generally describe what the ethical practitioner would do – because as you saw in our mashup table, there is a great deal of overlap. However, as identified in that table, there are activities that only ACM covers, so for the virtue framework, that would be the primary (or only) source for the prerequisite knowledge you bring to a virtue-ethics decision making framework involving those tasks.

The most important feature of our discussion of these resources is the fact that, once you have a bit of familiarity with them, you can either use them for your decision making, or recognize that they do/do not have key guidance. Recall that, "The entire community of scientists and engineers benefits from diverse, ongoing options to engage in conversations about the ethical dimensions of research and (practice)," (Kalichman, 2013: 13), and note that this is echoed in the preamble to the ACM CE: "Open discussions about ethical issues promote (this) accountability and transparency". When you *do* obtain guidance – but not the answer – in any situation, you can use that to help launch discussions about how "the ethical practitioner" responds when more than one principle is relevant in any case.

Considering whether the two frameworks suggest the same or different courses of action for the identified issue is one important feature for the making and justification of a decision, and also, reflection on that decision. More specifically, practice standard principles themselves may suggest different options and these must be recognized so that alternative actions can be formulated and evaluated. For example, ASA Principle C asserts that the ethical practitioner has responsibilities to science and to the client (as well as to the public, and possibly also a funder). This may arise similarly from ACM CE 2.5 (comprehensive evaluation of risks), where it is articulated that the responsibility is to provide objective evaluation (of a system) to the public as well as to clients. Sometimes these specific responsibilities may entail different decisions, if notifying the public of some result may benefit (or prevent harms to) the public while effectively divulging proprietary information which results in a net harm to the client. When such options arise (and conflict), open conversations with other ethical practitioners will be important for deciding what to do.

Chapter 1.6
Stakeholder analysis and the Utilitarian Decision-making Framework

We now turn to a new source of information, followed by a discussion of the use of all of the prerequisite knowledge in the ethical reasoning process. The new source of information for "prerequisite knowledge" includes a concise analysis of what you know to be harms and benefits that will impact stakeholders, the stakeholder analysis (SHA)[11].

A stakeholder is defined "one who is involved in or affected by a course of action" (Merriam-Webster); in the context of ethical case analysis, the stakeholder is simply an individual, or group, that might be affected by the outcome of the case. Clearly, identifying *who* might be affected, and the nature of that effect, are essential in understanding risks and benefits that might be associated with any decision or activity. Ethical case analysis implicitly requires that stakeholders are identified; however, this is not often a focus of instruction or practice. Because quantitative practice can have far-reaching implications (as described in both the ASA and ACM practice standards), consideration of stakeholders warrants more attention than is typical; so the stakeholder analysis template was created and appears in Table 1.6.1. The numbered notes (1-5) in the table are discussed on pp. 92.

Potential result: Stakeholder[1]:	HARM[5]	BENEFIT[5]	UNKNOWN[4]	UNKNOWABLE[3]
YOU[2,3]				
Your boss/client				
Unknown individuals[2]				
Employer				
Colleagues				
Profession				
Public/public trust				

Table 1.6.1. *Stakeholder Analysis template*

---

[11] This material on the stakeholder analysis is adapted from Tractenberg RE. (2019, April 23). Preprint. Teaching and Learning about ethical practice: The case analysis. Published in the Open Archive of the Social Sciences (SocArXiv), https://doi.org/10.31235/osf.io/58umw

There are two dimensions to the Stakeholder Analysis Template. The first dimension, captured in the columns, are *"Potential Results"*. These capture those effects of a decision or action, summarizing them according to whether or not they may represent net negatives. Potential negative results of any action or decision, or *harms*, include costing money, time, effort; negatively impacting reputations or persons; and other types of conceptual (intangible) or actual (tangible) damage. Like harms, potential positive results, or *benefits*, could be tangible or intangible – and they can have immediate or delayed effects. Benefits could include earning or gaining money; the removal of a harm; saving time or effort; improving reputation; and demonstrating expertise or superiority; among other things.

Since the effects of any action or decision may be negative for one entity, person, or group while positive for another, the potential result must be considered with respect to each *"Potential Stakeholder"* (the second dimension in the template). It may be surprising to realize that one of the potential stakeholders is *you*, the person making the decision. As described earlier, harms (costs in time, effort, and reputation) are easily recognized to yourself, as are benefits. These should be considered first because understanding the potential harms and benefits to yourself can also help you to recognize what they may be for the next stakeholder, your boss or client. If a decision costs *you* time (a harm), this could be a harm to your immediate *boss or cli*ent as well. By contrast, deciding to formally identify a data breach at work could be perceived as a harm to you (you could be punished if your boss does not want others in the company to find out) and a benefit to you (you would be forewarned that maybe your employer will soon be out of business or targeted by authorities). Only when the results of decisions or actions – in terms of both harms and benefits – are recognized can they be balanced against each other to make a *justifiable* decision (Tractenberg & FitzGerald, 2012). Note also here that if something represents a minor harm to you and a major harm to another stakeholder, or if it is a benefit to you but a harm to another stakeholder, then those are clearly *not exchangeable* –they represent different effects depending on the stakeholder. If anyone ever suggests to you that you should "simply treat others' data like you would treat your own", this should make you stop and think (rather than giving you a simple rule to follow): decisions you make that optimize benefits/minimize harms to *yourself* as the stakeholder are not necessarily going to also benefit/minimize harms to <u>others</u>.

You and your boss/client are fairly clear, recognizable, stakeholders. By contrast, *"unknown individuals"* are not recognizable stakeholders per se, but if you make a programming decision in the creation of an algorithm, or a distributional assumption in an analysis, there could be predictable but not-specific results for these unknown individuals. A simple example is males versus females: prior to 2017 the National Institutes of Health did not require that genetic sex effects should be considered and potentially modeled separately – making the scientific and statistical assumption that all human subjects (after controlling for age, height, and weight in many cases) are exchangeable[12]. The decision to require specification of sex effect hypotheses in proposed research, or to justify not including such specification, represents an acknowledgement that failing to consider sex effects in biomedical science is not appropriate, and is insufficiently rigorous. In terms of harms or benefits, failing to consider that heart attack symptoms differ for men and women (e.g., Coventry et al. 2011), and making assumptions in analyses or algorithms that *ignore the specific symptoms that women experience*, will end up potentially harming "unknown women". The benefit to "unknown men" might be "more is understood about the symptoms prior to a heart attack" –but because we (now) know that men and women differ, the corresponding benefit does not accrue to women. Medical professionals may never warn women about what might in fact occur to *them*. Thus, the assumption that "human subjects are exchangeable" has great potential to do tangible and lasting harm to those – unknown individuals - about whom such an assumption is wrong. Even when we know that "women experience heart attack symptoms differently from men", we will not know the specific women for whom the decision to model all humans the same will have this harm, thus the category of stakeholder is "unknown individuals". Unknown individuals also includes customers – e.g., all those customers who are known to have data/records associated with them, even though the specific customers whose data will be breached may be unknown.

Identifying your *"employer"* as a stakeholder may seem more straightforward, but "your employer" could be a person or the corporate entity that is your company's name. Harms and benefits can accrue to both; making your decision about notifying the authorities about a data breach could affect the corporate "employer" without affecting the CEO/owner. The more important aspect of this stakeholder is that harms and benefits could accrue to them based on *your* performance or decision. If you are self-employed, you are the face of your company; so while a benefit may accrue to *you* when you fail to notify the authorities about a data breach "at your company" (e.g., save yourself a lot of paperwork; limit the likelihood that a reporter will publicize the breach and name you specifically, etc.), this would actually constitute a harm to your

---

[12] See, e.g., https://www.drugabuse.gov/publications/research-reports/substance-use-in-women/importance-including-women-in-research

employer (your personal brand) – demonstrating that your company cannot be trusted by a potential customer.

Few practitioners in statistics and data science work alone, so recognizing that decisions we make about ethical problems can affect our *"colleagues"* should not be surprising. Some statisticians and data scientists are the only professional in that field on a team or in a working group – meaning that our colleagues in those situations are unaware of the ethical practice guidelines we are obliged to follow. If you fail to act on an ethical guideline principle, non-statisticians may never find out; but if they do, their trust in you would be diminished (a harm to you). You might, however, inadvertently create a situation where they might also be faced with ethical dilemmas caused by your action or decision; complicating their jobs creates a harm to *them*.

Professional Guidelines are developed by those in practice, in part, to delineate the qualifications to practice and to encourage public trust in *"the profession"* (Tractenberg et al. 2015). Some fields (e.g., medicine, law) have methods for controlling licenses to practice; all those who violate professional or ethical guidelines harm the profession – by suggesting that these controls do not act to keep "bad actors" out of practice. Those who ignore ethical challenges do not strengthen the profession – even if ignoring data breaches seems to benefit the individual ("keep your head down"; "go along to get along"), they harm the profession overall. When an individual, acting in a professional capacity, behaves unethically or ignores unethical behavior, the profession is impacted negatively.

Another stakeholder to consider is the *public/public trust*. When a representative of a profession is unethical, not only is that a harm to their profession but it also diminishes the public's trust in the profession, how that profession is regulated, and whether or not federal funds should be allocated to support or otherwise engage with that profession. The public – people who are not (yet) your customers, and people who are definitely "unknown individuals", but whose stake in the ethical practice of statistics and data science is very real – represent the cultural context in which we are educated, trained, and employed. The public also influences legislation –for or against our profession. Western culture tends to believe itself to be evidence-based: decisions are supposed to be supported by evidence and data (although see Pencheva, 2019 and McGoughy 2019, both discussing the campaign for Britain to leave the European Union ("Brexit"). Part of that evidence is "evidence that practitioners of statistics and data science can be trusted" – harms accrue to the public trust when, e.g., data breaches happen – the public may seek to require stricter legal controls on data collection (e.g., the General Data Protection Regulation, (EU)

2016/679; see https://en.wikipedia.org/wiki/General_Data_Protection_Regulation). Public sentiment towards federally funded research can also become negative if the public, or the public trust, are harmed. When practitioners act contrary to ethical practice standards, it detracts from "evidence that practitioners of statistics and data science can be trusted" – in fact, providing evidence against this argument.

It is also important for the person completing the stakeholder analysis template to recognize two types of unknown information, which are the final two columns in the table:

- *Unknown.* It is possible for a decision to be required early in a project (for example), before an effect for a given stakeholder can be established as a "harm" or a "benefit". Thought experiments, in which these effects are imagined - rather than observed or remembered from personal experience - (https://plato.stanford.edu/entries/thought-experiment/) or simulations can help to determine which of these (harm or benefit?) is more likely; and an important aspect of the stakeholder analysis template is to document where more information is needed. Additionally, whether a potential stakeholder is – or will become – an actual stakeholder may also need to be determined. Not all stakeholder analysis tables will include information in the "unknown" column, either because it simply cannot be guessed about (whether or not harms, benefits, both or neither may accrue at some future time) or they are actually completely predictable (so, are not "unknown").

- *Unknowable.* Both early and late in a project, it may be simply impossible to determine whether the effects for a particular stakeholder will be positive (benefit) or negative (harm). Whenever something appears in the "unknowable" column, it suggests that whatever decision is taken currently may need to be revisited in the future. Not all stakeholder analysis tables will include information in the "unknowable" column, either because harms and benefits are actually quite predictable (so, are not "unknowable"), or because the "unknowable" just comprises so very much that it doesn't really help your analysis to take it into account. Recognizing something as "unknowable" does not mean it should be ignored, but rather suggests that more thinking or more specification is required –possibly both.

Now that we have discussed every cell (row x column intersection) of the stakeholder analysis template, we can consider what information it might hold as a whole.

1. Knowing to whom harms may accrue can guide you to where the professional guidelines can assist in decision making. This is the topic of the next section.

2. Articulating the harms that may accrue to YOU is essential for you to "treat others' data as you would your own" (Loukides, Mason & Patil, 2018: Chapter 3). You need to recognize the harms that can accrue to you before you can compare those to you and those to others. Moreover, "others' data" could relate to your boss/client, your employer, unknowable others, or the public. Recognizing whether or not benefits or harms accrue to these different types of "others" is the only way for you to make a decision about how you want other people to treat your data: in someone else's table, you are the client, an unknowable other, or part of the public.

3. If there are no recognizable harms, and plausibly no "unknowable" harms <for which your decision would be responsible>, then there can be no conflict. It is really important to recognize whether something truly is unknowable or if it is actually something that can be known – but you just don't know it. The key words here are "recognizable" and "plausible" – your failure to recognize something doesn't mean it does not exist. And, beware of straw man[13] or red herring[14] harms!

4. If there are plausible harms (or benefits) that you cannot identify, but you believe/suspect may exist, then there is insufficient information for you to make a decision and you need more information. Recognizing this – instead of making an uninformed decision – is currently not part of the norm. Learning how to use this table and complete a case analysis is essential for enabling informed decisions about ethical challenges for current and future practitioners.

5. All harms are not the same; all the benefits are not the same; and harms and benefits are not exchangeable.

Discussion about stakeholders and harms/benefits that may result from any of the standard activities in statistics and data science can strengthen the learners' engagement with GLs, and also encourage consideration –and acceptance - of the responsibilities to practice ethically that the GLs describe (lest any of the harms befall any of the stakeholders).

---

[13] "Straw Man": defined as "an argument, claim or opponent that is invented in order to win or create an argument", Cambridge English Dictionary.
[14] "Red Herring": defined as "something that takes attention away from a more important subject", Cambridge English Dictionary.

The virtue decision making framework for stakeholder analysis

Decision making frameworks come from formal ethics to help ... make decisions. Both the ASA and ACM documents assert that they were drafted with the intention to support ethical decisions, and the ACM guidance is presented in a utilitarian perspective while the ASA's perspective maps to virtue ethics. The **virtue ethics** perspective can generally be summarized as, "what would the (ideal) ethical practitioner do in this situation?" The stakeholder analysis identifies harms and benefits that may be difficult to reconcile with what the ideal practitioner would do when faced with the tradeoffs that the SHA clearly lays out. Clearly, the ideal (most ethical) practitioner would want to minimize harms, and it is not clear whether the ideal practitioner (described by the ASA ethical guidelines, for example) would prioritize the minimization of harms *to the profession* – because they *are* the ideal professional practitioner after all – or if they would instead opt to prioritize the solution for the public trust or for colleagues. While it is not specific in the ASA GLs that the profession is the priority, it is reasonable to infer that harming the public trust (in the profession), or the public (of which even the ideal practitioner is a member), or colleagues can be subsumed under prioritizing the profession. That is, if the ideal practitioner follows the GLs and always acts to minimize harms to the profession, they will by default also minimize harms to the public, public trust, and colleagues.

The utilitarian decision-making framework for stakeholder analysis

The stakeholder analysis may seem most obviously supportive of a consequentialist or utilitarian approach because it provides structure to help quantify, in some ways, the balance between harms and benefits. This makes the **utilitarian** framework helpful to sort out the positive and negative effects of a decision on each of the stakeholders. The utilitarian perspective can generally be summarized as, "how can benefits be maximized while harms are minimized in this situation?" The ACM CE specifies that "Computing professionals' actions change the world. To act responsibly, they should reflect upon the wider impacts of their work, consistently supporting the public good." I.e., "the public good" is specifically identified as what should be prioritized. We already know that the rows of the SHA are the stakeholders, and the columns help us to identify harms and benefits. We also know that some harms are worse than others, and some benefits are "better" than others. So, to summarize the effects on a stakeholder, we would need to decide on a "scoring" approach so that a worse harm has a stronger, more negative weight on your decision and a better benefit should have a stronger, more positive weight on your decision.

Utilitarianism is well-suited to a quantification of the SHA results; we can simplify our classification of harms and benefits to be "major" and "minor", where "major" harm would include non-zero cost in terms of money or some other limitation or opportunity cost; and a "minor" harm would be a non-zero, but not substantial cost in terms of time; a "major" benefit would be earning or gaining money, or health, or other opportunities; while a "minor" benefit would be a savings of time or effort that you already need to commit at work. There was a method proposed in the 1970s (Campbell, 1976) and revised in 2019 (Tractenberg, 2019-a) where evidence for decision making can be summarized, and we will briefly explore how to utilize this method with the SHA. The method is called Degrees of Freedom Analysis. The basic method is to get the decisions in columns, and the features that the decision can affect -or be based on- are in rows. With two levels (major/minor) of harm and benefit, we can assign -1 and -0.5 "points" to major and minor harms; +0.5 and +1 to major and minor benefits, respectively, and "zero" to "no effect". This simple scoring method can be used to see quickly if one or another stakeholder has more harms, or more benefits, than another. It also serves to lay out features that can inform the decision. Minimization of harm – a focus in the ACM Code – can be easily visualized with the SHA: any column (decision) with positive "points" would be favored over columns with zeros or negatives.

The utilitarian perspective seeks to maximize benefits while minimizing harms, so you can see that the SHA table may support decision making using the utilitarian perspective. That does not mean that the only framework that benefits from the SHA is this one, because the virtue ethics position may also require that the "ideal" practitioner would rank order, according to their potential to limit harms, the guideline or code principles that are relevant in any given situation. Although we have seen that the ACM CE specifies that the responsible computing professional acts to minimize harm or the risks of harm (and so uses a fairly explicit utilitarian perspective), the ASA GLs are also consistent with the SHA approach, particularly when different GL principles suggest different decisions (e.g., prioritizing the public or the client when following GL Principle C). Whether decisions are driven by the ACM or ASA, or from a virtue or utilitarian perspective, the impacts of different decisions can be compared in terms of the harms and benefits that are identified using the SHA.

Recall again that, "The entire community of scientists and engineers benefits from diverse, ongoing options to engage in conversations about the ethical dimensions of research and (practice)," (Kalichman, 2013: 13). Note that "What do I do?" is not as effective at starting a conversation that will lead to a consideration of alternatives and the selection of a justifiable decision from among plausible, ethical, options (KSAs 4-5). By contrast, formulating the situation based on your prerequisite knowledge (KSA 1) "If I am completely transparent in my reporting, the public trust is strengthened, but my employer will lose the ability to patent the method/results. How can we balance these obligations?" – and you can go even further by adding KSA 2 as well: "If I am completely transparent in my reporting, the public trust is strengthened, and no harms accrue to the stakeholders except my employer. The harms to my employer are limited to them losing the ability to patent the method/results. The utilitarian perspective suggests that more people are harmed – as is the scientific enterprise – if the public trust is not prioritized; meanwhile, the virtue approach tells me that the ethical practitioner honors *both* obligations to their employer and to the public. How can we balance these obligations?" We are repeatedly invoking Kalichman's assertion to underscore how difficult it actually is to engage in these conversations. At this point, you can plainly see the difference between the kind of conversation you would have if you started with "Now what!?" rather than the beginnings of a reasoned argument that utilizes the GLs or CE. You may also see that learning how to reason ethically really is a learnable and improvable set of KSAs: If your current reasoning would be limited to asking, "Now what do I do?", then once you begin to develop KSA 1, and learn about the GLs, CE, and SHA, your reasoning is already stronger – and the likelihood of a useful conversation about the ethical dimension of your practice increases. When you add in KSA 2, that conversation can begin to include comparisons (which stakeholder harms/benefits should be prioritized? Or is it better to adhere strictly to the GLs/CE?). Two of the objectives of this book are: 1) to help you recognize the importance of KSAs 1-2; and 2) to help you engage in these important conversations. Making and justifying the decisions (KSA 5) about how best to practice ethically may or may not fall solely to you, which is why we discuss those mainly only in Section 3.

The information (prerequisite knowledge) that a "virtue" and a "utilitarian" perspective utilize may be different and yet still identify the same decision or option to be "preferred" or "most consistent with the framework". For example, a virtue approach might argue that, because both the ASA and ACM describe the ethical quantitative practitioner as "honest", all strengths and weaknesses of a technical solution to a problem should be transparently reported. By contrast, the utilitarian approach may identify that the only harms arising from failing to report fully and transparently actually accrue to the public trust and to unknown individuals (i.e., neither you nor your boss/colleagues/employer), but no real benefits accrue to anyone to offset those harms. Clearly the rationale for the decision "report honestly and transparently" differs depending on the framework, but the decision the frameworks lead to/recommend is the same. Then justification of the decision would naturally include the fact that both decision making frameworks support the same decision (report fully and transparently), although they do it differently. Thus double-justification could then feature in your reflection on the decision.

Considering whether the two frameworks suggest the same *or different* courses of action for the identified issue is one important feature for both the making and justification of a decision and also, reflection on that decision. This will become more apparent in Section 3 where we explore decisions in response to situations where "something is wrong" but you cannot be sure exactly what it is – without an analysis. For now, we simply note that even when contemplating how to execute your work, Guideline (ASA) and Code (ACM) principles may suggest different options (for Ethical Reasoning KSA 4), and these must be recognized so that alternative actions can be formulated (KSA 4) and evaluated (KSA 5). Again, your ability to engage in a conversation where choices like these (e.g., should we follow ASA or ACM?) is one of the main objectives of the book, and you can't do that without a very firm understanding of these two sets of practice standards.

The interrelationships between the materials you bring together as your prerequisite knowledge and the other features of the ethical decision-making process will become clearer as we delve more deeply into ethical reasoning, particularly in Section 3.

Chapter 1.7
Returning to Ethical Reasoning:
Summarizing KSAs 1 & 2[15]

So far, we have learned that Ethical reasoning involves six KSAs, and we have explored two: KSA 1, prerequisite knowledge and KSA 2, selecting a decision-making framework. These are interrelated, as you have seen, because a utilitarian decision-making framework requires that you are able to quantify, in some way, the harms and benefits that any act or decision will generate, while the virtue decision making framework requires a fairly comprehensive understanding of what "the ethical practitioner" would do.

As noted earlier, the practice standards from the ASA and ACM are clearly important both a) for informing your understanding of "what the ethical practitioner would do" *and* b) as essential prerequisite knowledge for ethical quantitative practice. We also discussed (in Chapter 1) the fact that norms are generally accepted behaviors according to practitioners in the community; as such, familiarity with these standards is a critical part of forming a professional identity as a quantitative practitioner, even if this is not the only job you have.

As we saw in earlier chapters, the ASA and ACM professional practice standards enable professional identity formation and promote a coherent view of the profession by the public and other stakeholders (Tractenberg et al. 2015; Tractenberg-2019-c). While quantitative practice involves many different tasks, concerns about data safety, privacy, and confidentiality – features of how data are collected, maintained/secured, and shared (or left vulnerable to unintended uses) - have become more visible, and more worrisome worldwide. In the face of growing concerns about safety, privacy, and unethical commercial uses of personal data, many groups have created tools and methodologies to "help". One such tool is the Data Science Ethics Checklist (DSEC http://deon.drivendata.org/). This tool was designed to support decision making, but was developed from the perspective that "the primary benefit of a

---

[15] This chapter includes material originally published as Tractenberg R.E. (2019, April 30). Strengthening the practice and profession of statistics and data science using ethical guidelines. Published in the Open Archive of the Social Sciences (SocArXiv), https://doi.org/10.31235/osf.io/58umw

checklist is ensuring that we don't overlook important work". There are 20 items on the Data Science Ethics Checklist, grouped into five areas.

A. Data Collection

A.1 Informed consent: If there are human subjects, have they given informed consent, where subjects affirmatively opt-in and have a clear understanding of the data uses to which they consent?

A.2 Collection bias: Have we considered sources of bias that could be introduced during data collection and survey design and taken steps to mitigate those?

A.3 Limit PII exposure: Have we considered ways to minimize exposure of personally identifiable information (PII) for example through anonymization or not collecting information that isn't relevant for analysis?

B. Data Storage

B.1 Data security: Do we have a plan to protect and secure data (e.g., encryption at rest and in transit, access controls on internal users and third parties, access logs, and up-to-date software)?

B.2 Right to be forgotten: Do we have a mechanism through which an individual can request their personal information be removed?

B.3 Data retention plan: Is there a schedule or plan to delete the data after it is no longer needed?

C. Analysis

C.1 Missing perspectives: Have we sought to address blind spots in the analysis through engagement with relevant stakeholders (e.g., checking assumptions and discussing implications with affected communities and subject matter experts)?

C.2 Dataset bias: Have we examined the data for possible sources of bias and taken steps to mitigate or address these biases (e.g., stereotype perpetuation, confirmation bias, imbalanced classes, or omitted confounding variables)?

C.3 Honest representation: Are our visualizations, summary statistics, and reports designed to honestly represent the underlying data?

C.4 Privacy in analysis: Have we ensured that data with PII are not used or displayed unless necessary for the analysis?

C.5 Auditability: Is the process of generating the analysis well documented and reproducible if we discover issues in the future?

D. Modeling

D.1 Proxy discrimination: Have we ensured that the model does not rely on variables or proxies for variables that are unfairly discriminatory?

D.2 Fairness across groups: Have we tested model results for fairness with respect to different affected groups (e.g., tested for disparate error rates)?

D.3 Metric selection: Have we considered the effects of optimizing for our defined metrics and considered additional metrics?

D.4 Explainability: Can we explain in understandable terms a decision the model made in cases where a justification is needed?

D.5 Communicate bias: Have we communicated the shortcomings, limitations, and biases of the model to relevant stakeholders in ways that can be generally understood?

E. Deployment

E.1 Redress: Have we discussed with our organization a plan for response if users are harmed by the results (e.g., how does the data science team evaluate these cases and update analysis and models to prevent future harm)?

E.2 Roll back: Is there a way to turn off or roll back the model in production if necessary?

E.3 Concept drift: Do we test and monitor for concept drift to ensure the model remains fair over time?

E.4 Unintended use: Have we taken steps to identify and prevent unintended uses and abuse of the model and do we have a plan to monitor these once the model is deployed?

Note that, in addition to data collection (A) and security (B), the DSEC suggests that the ethical quantitative practitioner is also concerned with what happens to the data after it has been collected and secured (ethically). Thus, while the DSEC emerged at least partly in response to the worldwide rise in worries about data safety, privacy, and confidentiality, it does go beyond just those topics. This echoes the ASA and ACM perspectives that there is more to ethical practice with data than privacy, confidentiality, and data safety.

The DSEC was developed in 2018 (inferred from the citations listed on the deon.org site) and as it is described on the site, this development was independent of the ACM Code of Ethics. Neither the ASA nor ACM explicitly recognize the historical importance of "codes of conduct" for establishing professional identity through professional standards or norms (see Tractenberg et al. (2015) for discussion of the evolution of both organizations' codes, and their role in establishing norms). The deon.org objectives with the checklist do actually include influencing the "norms" of ethical data science, but the group/site do not include any mention of professional identity. Importantly, both the ASA and ACM guidelines/code preambles make it explicit that they do intend to support ethical decision making by practitioners (not solely members of the profession or these societies). Similarly, the deon.org developers assert that checklists (like the DSEC) enable quick(-er) *ethical* decision-making -with the implication that anyone working with data should have this kind of skill set. Worth noting is that unintended uses of data- or the computing system you might be developing - is *not* explicitly considered in the DSEC checklist – only unintended uses and abuses of the model(s) that take the data and act on them. Rendering decisions, and taking other actions, are generally considered in E4. In order to assure a data contributor that unintended uses of their data are protected against, DSEC specifies that the quantitative practitioner must contemplate A1, specifically, "subjects affirmatively opt-in and have a clear *understanding of the data uses to which they consent*" (emphasis added). However, it is also important to note that the DSEC does not confer on the practitioner any responsibility (like ASA does in its Principles- especially D) or obligations (implied in ACM Code) to the data contributor. The DSEC simply asks if the practitioner has considered the items on the list. Recall, "The entire community of scientists and engineers benefits from diverse, ongoing options to engage in conversations about the ethical dimensions of research and (practice)," (Kalichman, 2013: 13). In this sense, rather than enabling ethical decision-making, it seems that the DSEC checklist might actually enable Kalichman's (elusive?) conversations about ethical practice, rather than decisions that are ethical. The fact that the DSEC encourages consideration, and maybe conversations about, ethical practice is a step in the right direction. Ethical reasoning requires additional effort – and that "the ethical quantitative practitioner" has, and accepts, obligations and responsibilities to others is an integral part of developing a professional identity by this practitioner.

Also, although both data collection and security are "covered" in the DSEC, and three other important aspects of a data-intensive or quantitative project (modeling, analysis, deployment) are also included, there *are* other key features of professional practice that fall outside of any kind of "project" that this checklist could be used for. For example, the ACM CE includes a section (with 7 items) focused on leadership, which entails different decisions specific to the role and quite distinct from those considered in the DSEC. You can imagine that a supervisor or team leader may need to task different members of the team with addressing different sections of the checklist – possibly for efficiency's sake. The ASA GLs also include (in Principle G) specific responsibilities for leaders, supervisors, and mentors in statistical practice, as well as organizational obligations (Appendix) to promote the kind of environment and culture that supports ethical practice. Our point in introducing the KSA about prerequisite knowledge as fully as we do in this book is that a tool like the DSEC can be useful, but should not supplant or interfere with professional identity development. "I use a checklist!" is not a professional identity - "I am an ethical quantitative practitioner!" *is.*

So, the DSEC can support you identifying and assessing your prerequisite knowledge (ER KSA #1): if you do not know how to determine if data were obtained with sufficiently informed consent (DSEC item A1), then you recognize a gap in your prerequisite knowledge and that this gap must be addressed before any ethical decisions can be made. However, the items on this DSEC Checklist are all yes/no questions. If you answer "no" to any of them, does it constitute an ethical issue? To determine whether or not it does, the ASA and ACM practice standards are essential, because ethical obligations are explicit – meaning that, if you answer "no" on a DSEC item, then whether and how it can lead to an ethical issue or problem (ER KSA #3) can be found in those standards. It bears repeating that both the ASA and ACM are emphatic that following their practice standards promote *ethical use* of methods and tools/techniques from statistics, data science, and computation – and these standards are not optional or otherwise dependent on the user's professional identity or membership in these fields or organizations.

Based on what you have learned about the ASA GLs, ACM CE, and the DSEC, we can explore general features of the interactions between the CE, GLs, DSEC, and each ER KSA:

1. **Identify and 'quantify' your prerequisite knowledge (ER KSA #1).** Before you have done anything and are still planning, you can use the ASA GLs, ACM CE – and even the DSEC – to plan (using the items on all of these documents to make sure that you know, or have access to, all the information needed to go on in an ethical fashion). For ASA GLs, Principle A ("Ethical statistical practice supports valid and prudent decision-making with appropriate methodology") suggests you need a full understanding of the data and the methods, so that whatever you end up planning will be valid, interpretable, and reproducible. For ACM CE, 1.2 states that you need to avoid harm. Looking just at ACM 1.2 and ASA A here shows that even valid, interpretable and reproducible work that causes harm will be unethical. So, ER KSA #1 actually goes fairly far towards preventing unethical behavior, or not

allowing decisions to do unethical things, when you bring the ASA and ACM guidance to bear. Just using the DSEC (or any checklist) does not address harm prevention, but rather, stipulates that the ethical data scientist plans a response to harms that may arise (DSEC E1) or unintended uses to which technology may be put (DSEC E4). Note that the GLs & CE describe how the ethical practitioner may avoid these harms, but apart from informing the practitioner that they do in fact have responsibilities to these stakeholders, these practice standards (like the DSEC) do not describe exactly how the address those harms or unintended uses. All of these sources leave that to the professional discretion of the practitioner.

2. **Identify decision-making frameworks (ER KSA #2).** The ASA GLs and ACM CE describe "the ethical practitioner". In formal terms, this qualifies both GLs and CE as "virtue ethics" decision making frameworks. When you choose or decide to behave or operate/practice in a way that is consistent with these frameworks, *you are* "the ethical practitioner" (so naturally, acting or deciding inconsistent with these frameworks makes that person unethical). When you are making decisions using these frameworks, you are generally going to be modeling your decisions according to "what would the ethical practitioner do in this case?" Another way to use these frameworks is an approach called "utilitarian". Formally, this framework can be helpful in identifying the positive and negative effects of a decision. For the utilitarian perspective, you need to have a firm idea of what those positive and negative effects would be *on*, and that can be very complex to conceptualize. The utilitarian perspective can generally be summarized as, "how can benefits be maximized while harms are minimized?" However, it is possible for a decision to maximize benefits to you, or your colleagues, while *not* minimizing harms to others; the opposite decision could minimize harms to others but not create any (or possibly even minimize) benefits to yourself and your colleagues (for more detail on harms and benefits and determining/prioritizing them, see Tractenberg 2019-d). Because a utilitarian perspective is so complicated, we focus here on the virtue approach, enabling us to just focus on what the "ethical practitioner" would do. This ER KSA (identify decision making framework) is absent from the DSEC by design (it is intended "to provoke conversations around issues"), whereas both ASA and ACM emphasize their intended uses in the professional's decision making.

3. **Identify or recognize the ethical issue (ER KSA #3).** What about what you are doing, planning, or have observed is *inconsistent* with the GLs or CE? What seems "questionable"? Amazingly, this is the obvious crux of any ethical decision, but is also incredibly difficult! If you look at the prerequisite knowledge you pulled together for ER KSA #1, you might see where an ethical issue (i.e., a potentially unethical decision, or a behavior someone else has done that is unethical) can arise or has arisen. Importantly, if you observe someone using statistics or data science in pursuit of unethical ends, they are violating the purpose of the ASA GLs, ASA GL Principle A, and ACM CE Principle 1.2. For you to *make a decision*, however – about what you should do (or not do), you need to go back to ER KSA #1 and make sure you identify all the ACM and ASA elements that you think are violated (or would be violated) by the specific behavior you are thinking of or observing. Note that these determinations are also required to identify harms (DSEC E.1) and unintended use/abuse (DSEC E4).

4. **Identify and evaluate alternative actions (on the ethical issue, ER KSA #4).** As noted, the GLs and CE are intended to support ethical decision making. That means there are at least two options, and you (the practitioner) must decide which one is most consistent with what the ethical practitioner would do in this case: do whatever is being asked, irrespective of what the ethical practice standards say, or change what you do so it is consistent with the practice standards. Here is an example where the DSEC could really help: choose the method/approach or whatever (depending on where you are in your project or workflow) that leads to a "yes" answer on the corresponding DSEC item. If you're deciding whether or not to obtain informed consent, or to provide all data contributors with "a clear understanding of the data uses to which they consent", you can tell (from the DSEC) that NOT doing these things leads to a "no" answer on that DSEC item (A.1) – meaning, if you do NOT do these things, then your data collection work will not be ethical. Every decision has to be between at least two things: it may be possible to abstract any choice you make in your work/workflow to something that matches what is on the DSEC. This would change a difficult-to-decide situation into more of a yes (I decide on something that creates a "yes" for the DSEC item) or no (I decide on something that creates a "no" for the DSEC item) decision. As we have discussed, "no" on a DSEC item suggests it is unethical. However, it might be impossible, or too costly (in time, effort, money, or some combination) to make a "yes" decision. The crucial thing is,

you have *now identified that you cannot make a "yes" decision – and you've also discovered why (because of costs).* Instead of saying to your boss or colleagues, "Hey! You can't do that!", you can say, "it looks like we need to X, but according to the ACM/ASA/DSEC, that could lead to unethical practice or charges that we are unethical. How can we work around this problem so we can accomplish X without doing it unethically?" What would the ethical practitioner do if s/he had to also do X? If there is no way for the ethical practitioner to do X, then that is a 3rd option: do not do X.

5.   **Note that**, if you had not made it this far in your ethical reasoning steps, you might have missed this – either done X without considering the negative effects, or done nothing because you guess X is maybe unethical but you're not sure, so you just skip it or ignore the fact that it might actually be doable *without* being unethical. The next person who needs to do X might not think this carefully about it, and might just do it, even though it is unethical. When you behave in ways that are not stewardly, it either fails to notify others that there is a risk of non-stewardly behavior, or it communicates that there is no reason to be stewardly or ethical. This is an example of how *your* ethical practice strengthens the whole profession: modeling stewardly behaviors opens that option up to observation by others, modeling unethical behaviors makes these appear to be options, which *weakens* the profession. Thus, the ASA Ethical Guidelines specify (Principle F) responsibilities to the entire profession. Specifically, F.5 states that the ethical statistical practitioner "Serves as an ambassador for statistical practice by promoting thoughtful choices about data acquisition, analytic procedures, and data structures among nonpractitioners and students."

6.   **Make and justify a decision (ER KSA #5).** At work, you make/will make hundreds of decisions all the time. Clearly you can't take the time to think this carefully about every single one of them. The point here is that you have identified an ethical problem (with your ER KSA#2), but not only did you identify it, you also came up with *three* alternatives (1. do it, ignore what the ethical practitioner would have done; 2. find a way to do it in a way that the ethical practitioner would also do; 3. Determine that there's no way for an ethical practitioner to do X – it can't be done ethically, so do not do X). You might actually be able to justify all three of these decisions, but clearly two of them are not consistent with the ethical decision-making framework you have chosen. You might think to yourself, "Hey, I can just go back to KSA#3 and choose a different decision-making framework so what I want to

do *is* consistent with that framework!" It is never acceptable to seek or create "loopholes" in the GLs (unless you also seek to fix/close them). If an unexpected ethical challenge arises, the steward seeks guidance, not exceptions, in the ethical GLs. To try and justify unethical or underspecified behaviors ("no one said I couldn't!") is unprofessional, unethical, and not stewardly. In fact, this language was added to the Purpose of the ASA GLs in 2022, "To justify unethical behaviors, or to exploit gaps in the guidelines, is unprofessional and inconsistent with these guidelines."

7. **Reflect on the decision (ER KSA #6).** Depending on the case, or the decision you walked through these KSAs to make, there might be a huge amount of reflection. You might re-think the decision periodically throughout your career – new information or technology may come along that reduces the costs associated with doing X ethically, for example. However, in this case we want to focus our attention on stewardship. Thinking about the next person who has to think through this same decision or a similar one, what additional information would be/would have been helpful to you (was there additional prerequisite knowledge you wish you had)? Thinking about your experience reasoning your way to an ethical decision, does a decision like the one you thought through help to create the culture that promotes fluency in ethical reasoning and/or a more ethical workplace? Obviously, the steward wants the next person who has to make this or a similar decision to think it through and choose the ethical, stewardly option. Reflecting on your decision and decision-making process may increase the likelihood of the next decision maker doing so in an ethical way.

Chapter 1.8
Aligning prior NIH/NSF training to promote ethical quantitative practice.

There is other knowledge that you may have obtained through standard university or college training in "responsible conduct of research" (or RCR), as this is required for any student or trainee whose training is funded wholly or in part by United States federal agencies (e.g., National Institutes of Health, NIH; National Science Foundation, NSF[16]). These US organizations define RCR as follows: "responsible conduct of research is defined as the practice of scientific investigation with integrity. It involves the awareness and application of established professional norms and ethical principles in the performance of all activities related to scientific research." Outside of the US, the focus is less on individual scientific conduct/misconduct and more on ensuring that research is driven by, and benefits, the public or social good (e.g., Kretser et al. 2019) and this is termed "responsible research and innovation" (RRI). Historically, RCR has tended to focus on ensuring that research and the scientific enterprise are engaged in with integrity (by the individual scientist, and supported by the research institution/organization) while RRI tends to focus on ensuring that whatever research is undertaken yields some social good. However, the 2017 consensus report, *"Fostering Integrity in Research"* from the US National Academies of Sciences, Engineering, and Medicine (NASEM) noted that, "Scientists are provided with opportunities and freedom to pursue new knowledge and train future scientists with the implicit understanding that they are responsible for the conduct of their research and the reliability of the knowledge they produce and that *they must conduct their research responsibly as a duty to the public.*" (Emphasis added). Moreover, this 2017 report identified a need for the research and scientific communities to focus energy on both the prevention of research misconduct and also what was originally called "questionable research practices" but is now called "detrimental research practices", because they have been recognized as not simply questionable but as actively damaging the public trust and the scientific enterprise.

---

[16] National Science Foundation. (2017). Training in responsible conduct of research—A reminder of the NSF requirement. https://www.nsf.gov/pubs/issuances/in140.jsp.

"The "responsible conduct of research" (RCR) comprises interactions with subjects (human and non-human), it also involves interactions with other scientists, the scientific community, the public, and in some contexts, research funders." (Tractenberg, 2016-b). This suggests that it is not a simple matter at all, contradicting the statement as recently as 2009, *"In general terms, responsible conduct in research is simply good citizenship applied to professional life."* https://ori.hhs.gov/content/preface. In her article identifying an issue and future challenge for ethics education in 2015, Professor Marie-Jo Theil commented, "ethics suffers from the idea that it can be practiced instinctively, without any learning, just by appealing to 'common sense'." (Theil, 2015: 45).

In addition to the extensive attention this book lavishes on the ethical practice standards for statistics and computing, there should be a sense in the reader that "common sense" or "simply good citizenship applied to professional life" are *not* accurate representations of either the relevance of ethical practice or the ease with which it should be developed. Owen, Macnaghten & Stilgoe (2012) note that "(s)cientists already have responsibilities, including those associated with the concepts of research integrity...RRI, however, confers new responsibilities." (p. 756). Owen et al (2012) argue that, in addition to the (complex) responsibilities that scientists have – which should be better understood by a reader having examined the ethical practice guidelines for two different disciplines – there are additional ones accruing when the relevance of innovation and research to the public good is considered. As you might suspect, this consideration is at least partly a driver of the ACM CE.

Information about what comprises RCR has been described using a list of topics deemed to be essential to "responsible conduct of research" by the National Institutes of Health (NIH) in the United States (see https://grants.nih.gov/ grants/guide/notice-files/NOT-OD-10-019.html). In January 2022, the RCR topics list was updated with some amendments, and some entirely new elements. Table 1.8.1 gives the updated list, with the additions and modifications indicated in bold (see updated list at https://grants.nih.gov /grants/guide/notice-files/NOT-OD-22-055.html). The original list of topics arealso discussed in National Academy of Sciences, National Academy of Engineering, and Institute of Medicine (1992) Steneck, 2009, and especially in Macrina (2014), but are not included in the *Fostering Research Integrity* report from the National Academies of Sciences, Engineering, and Medicine (NASEM 2017). These topics are considered briefly here, and the interested reader should consult the Macrina (2014) book because each of the original 2010 list of topics is much more fully discussed there, including cases for readers to work through on each of the topics in the list. Additionally, online resources have been marshalled by the Resources for Research Ethics Education (RREE) program that was funded at the University of California, San Diego (http://research-ethics.org/topics/overview/). Note that, since the NIH updated its topic list in February 2022 (https://grants.nih.gov/grants/guide/notice-files/NOT-OD-22-055.html), any of the resources, books, and other reports have not yet been updated to reflect the new topics list.

In this chapter, we only skim the RCR topics to ensure that readers who need to document for a funder that such information is included in their ethical training can incorporate these ideas into their own case analyses and responses to discussion questions throughout Sections 2 and 3 (see Table 1.8.1). However, because some of these topics bear quite specifically on international norms that all quantitative practitioners need to be aware of – conflicts of interest; the absolute requirement for informed consent for the collection of data; and the definition of "research" – are given some elaboration. These three topics in particular are relevant to our consideration of how ethical practice standards (from the ASA and ACM in particular) relate to our everyday dealings with data.

	Planning/ Designing Ch 2.2	Data collection/ munging/ wrangling Ch 2.3	Analysis (perform or program to perform) Ch 2.4	Interpretation Ch 2.5	Documenting your work Ch 2.6	Reporting your results/ communication Ch 2.7	Engaging in team science/ team work Ch 2.8
conflict of interest – personal, professional, and financial – and **conflict of commitment, in allocating time, effort, or other research resources**			x			x	
policies regarding human subjects, live vertebrate animal subjects in research, and safe laboratory practices	x	x					
mentor/mentee responsibilities and relationships					x	x	
collaborative research, including collaborations with industry **and investigators and institutions in other countries**							
peer review, including the **responsibility for maintaining confidentiality and security in peer review**					x	x	

	Planning/ Designing Ch 2.2	Data collection/ munging/ wrangling Ch 2.3	Analysis (perform or program to perform) Ch 2.4	Interpretation Ch 2.5	Documenting your work Ch 2.6	Reporting your results/ communication Ch 2.7	Engaging in team science/ team work Ch 2.8
data acquisition and analysis; laboratory tools (e.g., tools for analyzing data and creating or working with digital images); recordkeeping practices, including methods such as electronic laboratory notebooks	×	×					
secure and ethical data use; data confidentiality, management, sharing, and ownership	×	×					
research misconduct and policies for handling misconduct				×			×
responsible authorship and publication				×		×	
the scientist as a responsible member of society			×				×

NB: For 2022, the NIH has added a new topic, "safe research environments (e.g., those that promote inclusion and are free of sexual, racial, ethnic, disability and other forms of discriminatory harassment)". This RCR topic is not addressed in this book because, while critical to ethical statistics and data science, it is beyond the scope of this book.

Table 1.8.1 *NIH RCR training topics and discussion questions in Section 3 by chapter/task*

Relating to the NIH topics list shown in Table 1.8.1, definitions and brief discussions follow.

Conflict of interest – personal, professional, and financial – and conflict of commitment, in allocating time, effort, or other research resources

A conflict of interest is defined as "a situation in which a person is in a position to derive personal benefit from actions or decisions made in their official capacity."

(OED https://en.oxforddictionaries.com) The Canadian Responsible Conduct of Research Office elaborates: "A conflict of interest may arise when activities or situations place an individual in a real, potential or perceived conflict between the duties or responsibilities related to research, and personal, institutional or other interests. These interests include, but are not limited to, business, commercial or financial interests pertaining to the individual, their family members, friends, or their former, current or prospective professional associates." (Based on the second edition of the Tri-Council Policy Statement: Ethical Conduct of Research Involving Humans [TCPS 2] Chapter 7).

Conflicts of interest (COI) must be identified and managed. In some cases, the identification of a COI requires the individual with the conflict to be recused or removed from the project or task for which the conflict arises; an example is in the provision of peer review of grants or papers, individual reviewers who work at the same institution (e.g., university) as the applicant or author would not be allowed to provide a review. In special circumstances, where there is highly limited expertise in an area and the only expert who can review the grant or manuscript happens to be at the same institution, the organizers of the review (the grantor or the journal considering the manuscript) will manage the conflict, and allow the person with the identified COI to contribute. Essential to note here is that it is not the individual with the COI who makes the decision. Management of the COI means that, while an individual may have interests that conflict, these must be transparently disclosed and also mitigated. Decisions that are made by an individual with a perceived or actual conflict of interest will always appear to have been driven by their own interests, rather than those of the project or others. Mitigating the conflict often entails delegating the decisions where the COI could, or even only appears to (but doesn't actually) affect the outcome. When decisions cannot be delegated, transparent and complete documentation of the decision-making procedure must be provided. Examples of decisions that cannot be delegated would be those that must be made by the sole quantitative practitioner on a project, with unique expertise and experience. Seeking guidance from organizational resources is strongly recommended whenever COIs arise or appear to arise.

Policies regarding human subjects, live vertebrate animal subjects in research, and safe laboratory practices

> NOTE that the US policy about human subjects as represented in the Code of Federal Regulations (CFR) changed, effective 19 July 2018 in the US, https://www.hhs.gov/ohrp/regulations-and-policy/regulations/terminology/index.html

These policies are quite extensive and require attention that is beyond the scope of this chapter and this book. However, because data collection is such a fundamental component of both statistical and computing practice in modern society, it is worth dedicating some time and energy to what it means to notify a human of what data you plan to collect, and to obtain their consent to do this. We focus on human subjects, recognizing that ethical treatment of animals in research is a critical consideration, because much of computing machinery, data science, and statistical practice - whether in business or science - revolves around proliferating methods for collecting data from humans. Note that ACM is silent on the source of data as a rule for being ethical (or not considering the data donors), whereas the ASA has Principle D specifically focused on "Responsibilities to research subjects, data subjects, or those directly affected by statistical practices, Data Subjects, or Those Directly Affected by Statistical Practices".

Sometimes called "the common rule" but officially called Title 45 CFR part 46 ("45 CFR 46") and comprising five subparts, (A-E), 45 CFR 46[17] describes how human research participants must be treated - and how their explicit and informed consent must be obtained before you treat them any way at all - when federal (US) funds are used to support research. Since many research institutions in the US receive federal funds for their research, most if not all research at these institutions (even that funded by foundations or other non-federal sources) follows 45 CFR 46. Part 46 is entitled "Protection of Human Subjects". Subpart A of CFR 46 presents the "basic HHS Policy" relating to human subjects' participation in research, with subsequent subparts of the policy adding protections for pregnant women, foetuses, and neonates (Subpart B); prisoners (Subpart C); and children (Subpart D). Subpart E focuses on the registration of institutional review boards (IRBs).

---

[17] https://www.hhs.gov/ohrp/regulations-and-policy/regulations/45-cfr-46/index.html

The contents of 45 CFR 46 A are extensive, and should be evaluated periodically – not because it changes often but because different parts of the code may apply to different research throughout a career. It is essential to note that, although many business applications of statistics and data science do not collect data in the same way that 45 CFR 46 Subpart A discusses (and was created specifically to address and guide), these rules and definitions apply widely in the US and internationally.

(e)(1) *Human subject* means a living individual about whom an investigator (whether professional or student) conducting research:

(i) Obtains information or biospecimens through intervention or interaction with the individual, and uses, studies, or analyzes the information or biospecimens; or

(ii) Obtains, uses, studies, analyzes, or generates identifiable private information or identifiable biospecimens.

Other important definitions include:

(4) *Private information* includes information about behavior that occurs in a context in which an individual can reasonably expect that no observation or recording is taking place, and information that has been provided for specific purposes by an individual and that the individual can reasonably expect will not be made public (*e.g.*, a medical record).

(5) *Identifiable private information* is private information for which the identity of the subject is or may readily be ascertained by the investigator or associated with the information.

(6) *An identifiable biospecimen* is a biospecimen for which the identity of the subject is or may readily be ascertained by the investigator or associated with the biospecimen.

And it is important to note that there are plans to regularly revisit this policy specifically as technology changes:

(7) Federal departments or agencies implementing this policy shall:

(i) Upon consultation with appropriate experts (including experts in data matching and re-identification), reexamine the meaning of "identifiable private information," as defined in paragraph (e)(5) of this section, and "identifiable biospecimen," as defined in paragraph (e)(6) of this section. This reexamination shall take place within 1 year and regularly thereafter (at least every 4 years). This process will be conducted by collaboration among the Federal departments and agencies implementing this policy. If appropriate and permitted by law, such Federal departments and agencies may alter the interpretation of these terms, including through the use of guidance.

(ii) Upon consultation with appropriate experts, assess whether there are analytic technologies or techniques that should be considered by investigators to generate "identifiable private information," as defined in paragraph (e)(5) of this section, or an "identifiable biospecimen," as defined in paragraph (e)(6) of this section. This assessment shall take place within 1 year and regularly thereafter (at least every 4 years). This process will be conducted by collaboration among the Federal departments and agencies implementing this policy. Any such technologies or techniques will be included on a list of technologies or techniques that produce identifiable private information or identifiable biospecimens. This list will be published in the FEDERAL REGISTER after notice and an opportunity for public comment. The Secretary, HHS, shall maintain the list on a publicly accessible Web site.

Finally, we see the definitions for *research*, which are essential for many scientists (even those working at private companies), but which may not necessarily apply for those working in industries that do not seek to "develop or contribute to generalizable knowledge":

(l) *Research* means a systematic investigation, including research development, testing, and evaluation, designed to develop or contribute to generalizable knowledge. Activities that meet this definition constitute research for purposes of this policy, whether or not they are conducted or supported under a program that is considered research for other purposes. For example, some demonstration and service programs may include research activities. For purposes of this part, the following activities are deemed not to be research:

(1) Scholarly and journalistic activities (*e.g.*, oral history, journalism, biography, literary criticism, legal research, and historical scholarship), including the collection and use of information, that focus directly on the specific individuals about whom the information is collected.

(2) Public health surveillance activities, including the collection and testing of information or biospecimens, conducted, supported, requested, ordered, required, or authorized by a public health authority. Such activities are limited to those necessary to allow a public health authority to identify, monitor, assess, or investigate potential public health signals, onsets of disease outbreaks, or conditions of public health importance (including trends, signals, risk factors, patterns in diseases, or increases in injuries from using consumer products). Such activities include those associated with providing timely situational awareness and priority setting during the

course of an event or crisis that threatens public health (including natural or man-made disasters).

(3) Collection and analysis of information, biospecimens, or records by or for a criminal justice agency for activities authorized by law or court order solely for criminal justice or criminal investigative purposes.

(4) Authorized operational activities (as determined by each agency) in support of intelligence, homeland security, defense, or other national security missions.

Although the US Federal regulations are currently specific to "research" involving "human subjects" as defined in the foregoing, the reader should be impressed by the concept *that all human data should be collected in ways that respect the data contributor's autonomy,* and recognizing their fundamental human and civil rights to determine whether or not to contribute data, based on correct and timely provision by the collector of the intended purposes of that data collection. The ASA and ACM practice standards include specific considerations that the ethical practitioner owes to research subjects and to those from whom data are collected. Because of their historical role in experimental research, the responsibilities of all those following the ASA GLs are described using terms that are quite consistent with the Federal Regulations, especially Principle D: "The ethical statistical practitioner does not misuse or condone the misuse of data. They protect and respect the rights and interests of human and animal subjects. These responsibilities extend to those who will be directly affected by statistical practices.". The ACM CE specifically outlines the ethical computing professional's obligations to respect the autonomy of those whose data are collected, which may or may not be interpreted as suggesting that animal data should be covered by this obligation (depending on your personal or contextual perspective on animals). Both of these practice standards are consistent with the 45 CFR 46 ideas that the individual or organization who/that collects –or designs systems to collect - data owes an obligation of respect for the rights of the individual who contributes that data.

Another important thing to note is that, because so much of current research is actually guided (regulated) by 45 CFR 46, these rules do in fact represent worldwide "norms" for ethical behavior relating to human subjects and obtaining consent from them to collect and utilize their data. Thus, even for quantitative practitioners who do not work in "research institutions", or who work for other organizations that collect and analyze data, but do not meet the 45 CFR 46 definition(s) of research, respect for the autonomy of humans to be informed of, allow or disallow, and to direct the use of their personal data is in fact globally acknowledged. Not all institutions or organizations recognize these rights, but the ASA and ACM practice standards charge all those whose activities they seek to guide to respect this autonomy.

Mentor/Mentee

Mentor (An experienced person in a company or educational institution who trains and counsels new employees or students)/**mentee** (A person who is advised, trained, or counselled by a mentor) **responsibilities and relationships**

https://oir.nih.gov/sites/default/files/uploads/sourcebook/documents/mentoring/guide-training_and_mentoring-10-08.pdf

There are important and specific requirements and responsibilities for those who offer mentorship - advice, guidance, and other professional support – to those who are more junior or actual trainees in any organization. The mentor is an experienced individual – oftentimes, but not necessarily, older than the mentee (the one being trained or supported). Mentor/ Mentee responsibilities and relationships have a special place in the NIH topics list because one way that expertise in science is recognized by these bodies is by the successful mentorship of new members of the scientific community. Thus, to be designated a "mentor" is an honor, and should represent acknowledgement of the mentor's experience, expertise, and (perhaps most importantly) potential to support the development of new professionals in the field. To be designated a "mentee" should confer a commitment to the mentee's professional growth while also designating that mentee as similarly committed to their own growth and development within the field. The ACM practice guidelines do not separate mentorship from leadership or other aspects of ensuring that the entire profession benefits from, and is not harmed by, their activities. Both ASA and ACM standards also specify an obligation to be respectful in all interactions, refraining from –and actively discouraging – intimidation, harassment, bullying and other disrespectful and discriminating behaviors. The ASA GLs specify that the ethical statistical practitioner –including all those who use statistics (see Preamble) - has a responsibility to not only refrain from these behaviors but to act to intervene or stop them when this is possible; the ACM CE specify that all those who use computing in an impactful way (see preamble) have an obligation to follow and encourage adherence to the CE. ASA went further in 2022, charging leaders, supervisors, and mentors of statistical practice with specific expectations (Principle G.1) to "Ensure appropriate statistical practice that is consistent with these guidelines. Protect the statistical practitioners who comply with these guidelines and advocate for a culture that supports ethical statistical practice." Thus, both sets of practice standards offer specific guidance for the mentor, and protections to the mentee, in terms of prevention of unprofessional and disrespectful behaviors. The ASA GL Appendix also outline "responsibilities to use statistical practice in ways that are consistent with these guidelines, as well as to promote ethical statistical practice" for organizations and institutions.

Collaborative research including collaborations with industry and investigators and institutions in other countries

This item appears on the topics list because it originates in a US Federal funder that most commonly engages with universities and other institutions of higher education (National Institutes of Health), rather than with industry or business. As the reader will readily comprehend,

a.   collaborative research happens within academia as well as across the boundaries of academia and industry;

b.   collaborative work involving data *that does not qualify as "research"* happens within academia, outside of academia, and across the boundaries of academia and industry;

c.   whether your role is "collaborator", "team member", or "employee", you are probably engaged in "collaborative" work that will have ethical practice dimensions; and

d.   all of this collaborative work may involve data – and so, ethical practice of data-centered activities is an essential aspect of engagement whether or not it qualifies as research.

An important consideration is that the individual(s) responsible for data-centered or quantitative work in a collaborative project have responsibilities outlined by the ASA GLs and ACM CE, but working on a team (whether it's research or not; whether it's truly collaborative or not; whether it is within academia, within industry, or across these boundaries) is also discussed in the ASA GLs from the practitioner's perspective (Principles E and F) and from the perspectives of leaders (Principle G) and organizations/institutions (Appendix). Note that there are no cases directly related to "collaborative research" in Table 1.8.1; this is because the ASA and ACM practice standards do not differ depending on collaborative partners. Any of the cases can be reframed for discussions (or, your reasoning) about collaboration specifically

Peer review

Like with "collaborative research", this item is specific to the origins of the list being generally academic and exclusively focused on research. However, as is pointed out in Chapter 2.4, there are many different roles for the review or evaluation of peer work. The NIH topics list refers to the ubiquitous "peer review" of research ahead of publication, which is a form of vetting of new work by experienced scientists ideally to ensure that contributions to the base of knowledge are meaningful. "Review" by peers should be construed more broadly than this when considering statistics and data science – because the evaluation (review) of peer work should be done in an ethical way, so that contributions to the scientific and practice knowledge bases are both meaningful and ethically done. The ASA GLs include guidance on evaluation of the analysis of data, and in some cases to an analysis of how people use the data or of others' use of data, results, or methods – whether these are for research or not. Moreover, the ASA GL (Principle E and F, G and Appendix) highlight how important the practitioner's interactions with others, including peer review (Principle F.2) should be transparent, reproducible, rigorous, and respectful. The ACM CE also specifically addresses the evaluation of systems or their plans/designs, and to the analysis of activities, specifically with respect to risks in terms of the effects of computing on stakeholders.

Data acquisition and analysis

Secure and ethical data use; data confidentiality, management, sharing, and ownership

Responsibilities pertaining to acquiring and managing data have been on the NIH topics list since its earliest versions, with the ethical implications for "data acquisition" being a core and focal part of all training in "responsible conduct of research" since this idea was first integrated into federal funding programs. That is, ethical data acquisition was originally about ensuring that data contributors, specifically research participants but soon grew to encompass animal research subjects. The importance of data management, sharing, and ownership has completely changed in the past 10 or so years, with data sharing becoming a very large part of data management – because the sharing has to be done in a way that protects the privacy and confidentiality of the data contributors, and this has implications for data management. Moreover, current work with data involves far more than just "research", and "data acquisition" goes beyond experimental contexts. Therefore, this topic is almost redundant with the majority of the ASA Ethical Guidelines (all versions), and has considerable overlap with the ACM CE as well. However, both the ACM and ASA outline responsibilities relating to these core NIH topics in actionable and defensible ways.

Research misconduct and policies for handling misconduct

In December 2000 the US Office of Science and Technology Policy defined "research misconduct" as "fabrication, falsification, or plagiarism in proposing, performing, or reviewing research, or in reporting research results" (Steneck 2009 – Ch 2 (ONLINE, https://ori.hhs.gov/content/chapter-2-research-misconduct-federal-policies)

However, particularly for our purposes in this book, the specificity of "misconduct" to explicitly pertain to research is quite limiting- it suggests that the only "bad" behavior we need to prevent is that which represents "fabrication, falsification, or plagiarism" – also known as "FFP" – which is an absurd limitation – and which also ignores the use of ethically obtained data for not-permitted uses, or the unethical gathering of data. You can see from the ASA and ACM practice standards that FFP is only a tiny proportion of the potentially worrisome behaviors that all ethical practitioners need to be aware of, and guard against. Instead, a more relevant view was published in 2017 by the National Academies of Sciences and Engineering and the Institute of Medicine (NASEM):

> "Much of the discussion, thinking, and actions aimed at fostering research integrity has revolved around the actions of miscreant individuals in committing acts of research misconduct and its components—fabrication, falsification, and plagiarism. Actions that *Responsible Science* [a 1992 report by NASEM] characterized as questionable research practices have received less attention. The accumulation of knowledge has brought the critical need to address these elements to the fore. *Actions such as failing to retain or share data and code supporting published work in accordance with disciplinary standards, practices such as honorary or ghost authorship, and using inappropriate statistical or other methods of measurement and data presentation to enhance the significance of research findings are clearly detrimental to the research process and may impose comparable or even greater costs on the research enterprise than those arising from research misconduct.*" (NASEM 2017, p. 206, emphasis added).

This perspective has not yet (2022) been integrated into the NIH resources (https://oir.nih.gov/sourcebook/ethical-conduct/research-ethics/nih-policies/investigation-allegations-research-misconduct) - nor into the topics list revised in 2022- but is extremely important for our thinking about "norms" and how important it is for all new practitioners to bring at least some preparedness to reason ethically into the workplace. That is, rather than simply focusing our attention on preventing FFP, all scientists, and those who use scientific tools and techniques (such as quantitative practitioners, statisticians, and computing professionals) need to understand the devastating negative impacts that "questionable research practices" can have. The NASEM report renamed these "detrimental research practices" because there is literally no question that these should not be engaged in (i.e., they're not actually "questionable" – they're frankly unethical), and more to the point, these behaviors, and those who engage in them, are undermining the scientific enterprise. Whether you are, or plan to be, a scientist, the quantitative practitioner is using scientific principles in some, if not most, of their daily work. We focus on detrimental research practices here not only because science and the scientific community deserve better, but because whenever practitioners engage in questionable or detrimental practices, they act to undermine the integrity of their profession and the reputation of others in that field. Whenever such practices spread – because people observe them and think, "hm, that could be the new norm" – real harms accrue. You may notice that both ASA and ACM include language intended to discourage detrimental practices that utilize statistical and computing technology or methods, particularly as these practices are used by those who may not consider themselves obliged to follow the ASA or ACM standards. Both standards charge practitioners and organization members alike to promote ethical practice among those in their field, students, *and* those who use the techniques/methods or technology "in an impactful way" (ACM Preamble) or "in their professional work" (ASA Preamble).

Thus, while the NIH continues (in 2022) to focus on preventing "research misconduct", the scientific community is moving on to preventing "detrimental research practices", and the ASA and ACM in particular seek to promote *ethical practice* by all who utilize their methods/technology.

Responsible authorship and publication

While its importance for professionals outside of university or academic settings may be negligible, authorship and publication are the core professional activities of the academic scientist. Because publication is so essential to "success" in academia, practices surrounding publication are often subject to "questionable or detrimental practices", and this has begun to attract increasing attention over time. In response to these practices, NASEM (2017) made the following recommendation regarding authorship specifically:

> **Recommendation Five**: Societies and journals should develop clear disciplinary authorship standards. Standards should be based on the principle that those who have made a significant intellectual contribution are authors. Significant intellectual contributions can be made in the design or conceptualization of a study, the conduct of research, the analysis or

interpretation of data, or the drafting or revising of a manuscript for intellectual content. Those who engage in these activities should be designated as authors of the reported work, and all authors should approve the final manuscript. In addition to specifying all authors, standards should (1) provide for the identification of one or more authors who assume responsibility for the entire work, (2) require disclosure of all author roles and contributions, and (3) specify that gift or honorary authorship, coercive authorship, ghost authorship, and omitting authors who have met the articulated standards are always unacceptable. Societies and journals should work expeditiously to develop such standards in disciplines that do not already have them.

The fact that computing professionals do not always work in academic settings may be reflected only indirectly in the ACM CE (i.e., only mentioned in ACM 1.5 as a responsibility for computing professionals to respect "authors' works"), while authorship is a far more common aspect of the statistician's role (and is discussed briefly in ASA A.5, A.6). The ASA GLs are replete with allusions to writing, if scholarly publications are not the objective, such that transparency, rigor, reproducibility, and validity in all reporting are repeatedly encouraged. Neither practice standard follows NASEM Recommendation 5 (although the author of this book has recommended to the ASA that the next revision of the GLs should incorporate the definition below). Since 2004, an **author** has been defined by the International Committee of Medical Journal Editors (ICMJE) as an individual who meets the following four criteria:

- Substantial contributions to the conception or design of the work; or the acquisition, analysis, or interpretation of data for the work; **AND**
- Drafting the work or revising it critically for important intellectual content; **AND**
- Final approval of the version to be published; **AND**
- Agreement to be accountable for all aspects of the work in ensuring that questions related to the accuracy or integrity of any part of the work are appropriately investigated and resolved.

http://www.icmje.org/recommendations/browse/roles-and-responsibilities/defining-the-role-of-authors-and-contributors.html

The ICMJE criteria are interesting from a variety of perspectives. They were derived by editors of medical journals, rather than, for example, by consensus among all authors of their respective journals. Like the ASA and ACM practice standards, which were created by- and whose maintenance is charged to – a small subset of the much wider community on their respective committees for ethical practice ("professional ethics"). The ICMJE criteria represent what experienced practitioners have determined to be the "norms". That is, these do not represent a consensus of "what *is* done" but rather, "what *should be* done". The ICMJE authorship criteria listed above were published originally in a form similar to the final four criteria in 2004[18]. The two criteria listed last are possibly the most important ones: the first two (substantial contributions to the conception/design or acquisition/analysis/interpretation of data; and drafting or revising the paper are straightforward representations of concrete contributions without which a manuscript would be very different (if not impossible) to generate. When a researcher seeks to answer a scientific question, they often must partner with quantitative collaborators (statisticians, data scientists) for the design and collection, as well as analysis and interpretation, of data that can answer the question. The quantitative practitioner will have little difficulty in demonstrating that they meet these two criteria – and should ensure that their interest in fulfilling these roles, and availability to do so, are documented.

However, having "final approval" of what will be published means literally that if the quantitative practitioner does not approve of the contents, then the work would not be published. Moreover, the final criterion, "(a)greement to be accountable for all aspects of the work", can often be the most problematic for the quantitative practitioner. Ensuring that all statistical and computing work is correctly described (documented and interpreted), and that all conclusions are supported by the analyses and methods that were used, is perhaps uniquely the role of the quantitative team member(s). It is crucial that these members recognize, and require that other team members also recognize, their obligations when it comes to the transparent, valid, and accountable reporting of work that features quantitative and/or computational methods. Integrating the ICMJE authorship criteria into your own professional identity would mean that you would not participate in projects – or allow your name to appear as a co-author on work – where you were not allowed to fully realize these four obligations. Thus, augmenting the ASA and/or ACM GL/CE with the ICMJE criteria for authorship would both satisfy the NASEM recommendation, and would also be consistent with both standards' stipulations that ethical practitioners should seek to encourage others to follow their ethical guidelines, particularly when reporting work that is contingent on statistical, computational, or otherwise quantitative work.

---

[18] http://www.icmje.org/recommendations/archives/2004_urm.pdf

The scientist as a responsible member of society

The final topic on the NIH list is a bit of a catch-all. Both the ASA and ACM recognize the social impact and importance of the work supported or done by statistics, computing, and data science, and the contributions that their techniques and technologies do, or have the potential to, make to society and the public good. The ACM focuses its members' attention explicitly on the public good and the avoidance of harms in many different places throughout its guidelines. Moreover, the ASA GLs include a specific principle (C) with five elements to capture "Responsibilities to Stakeholders", with a separate principle outlining responsibilities to the humans and animals from whom data may be obtained (D, "Responsibilities to research subjects, data subjects, or those directly affected by statistical practices, Data Subjects, or Those Directly Affected by Statistical Practices"). Both practice standards, then, recognize responsibilities of their respective practitioners to treat data, and the rights of those who contribute data, ethically. They both also recognize that the social context, within which their techniques and technologies are applied, is relevant for ethical decision making and practice. Importantly, because both organizations (ACM, ASA) maintain their ethical practice standards in a dynamic and ongoing way, they strive to address both the contemporary ethical issues and those that are more longstanding – and seek to support their respective practitioners' ethical decision making in all ethical issues, including those that are recognizable today and those that might arise in the future.

Questions for discussion:

The NIH promulgates their list of key topics (published in 2009 with a description that they represented (at that time) 20 years' worth of delivering training in RCR (https://grants.nih.gov/grants/guide/notice-files/NOT-OD-10-019.html). As discussed, the topics list was updated in February 2022 (https://grants.nih.gov/grants/guide/notice-files/NOT-OD-22-055). They note specifically that "there are no specific curricular requirements for instruction in responsible conduct of research" but essentially dictate the content of "acceptable" RCR training to comprise the list of topics shown in Table 1.8.1.

The 2009 NIH policy (https://grants.nih.gov/grants/guide/notice-files/NOT-OD-10-019.html) states, "While courses related to professional ethics, ethical issues in clinical research, or research involving vertebrate animals may form a part of instruction in responsible conduct of research, *they generally are not sufficient to cover all of the above topics.*" (Emphasis added). This policy was not updated in 2022.

1. Discuss whether a course emphasizing professional ethics relating to computing and statistical practice using the ACM and ASA practice standards would or should be considered "acceptable"; what would be lacking from such a course, and can you derive a more comprehensive list of topics that is relevant to your ethical practice and the development of a professional identity?

Consider the 2015 European Group on Ethics in Science and New Technologies (EGE), which published their General Activity Report, 2011-2016. In their Statement on Research Integrity, the 'European Code of Conduct for Research Integrity', described the characteristics of the ethical scientist: "honesty; reliability; objectivity; impartiality and independence; open communication; duty of care; fairness; and responsibility for future science generations." https://ec.europa.eu/research/ege/pdf/ege_genral-acivity-report_2018.pdf P. 32.

2. Discuss the alignment –or lack thereof – between the EGE characteristics, the RCR topics list, and the content of either the ASA or ACM practice standards. How can ASA or ACM documents be used to support the training of new scientists that meet the EGE characteristics? Do you think the additional topic areas added to the NIH list in 2022 are helpful?

Chapter 1.9
Identify or recognize the ethical issue: KSA 3

We continue to explore all of the KSAs of ethical reasoning in Section 1, and since we grouped KSAs 1 (prerequisite knowledge) and 2 (decision making framework), we seem to be bouncing around these two KSAs. In fact, this underscores the fact that Ethical Reasoning is not usually *linear*, leading directly from the first KSA through all the others, and then to the final "correct" answer. Instead, recognizing when you have assembled sufficient prerequisite knowledge (KSA 1) to both identify (KSA 3) and make a decision that will address the problem (KSA 4) can benefit from organizing your knowledge according to the decision-making framework (KSA 2).

As you may suspect –given that most of Section 2 is devoted to just these - there is a lot of material that makes up "prerequisite knowledge". We noted earlier that simply knowing the GLs or CE is not sufficient to make and justify ethical decisions throughout quantitative practice, and in fact KSA #3, identifying or recognizing the ethical issue, is one of the more difficult parts of making decisions ethically. Gunaratna & Tractenberg (2016) discussed how to use the ASA GLs to identify what the actual ethical problem is/problems are. In this chapter we revisit this earlier material, and also expand it to show how both the ACM CE and stakeholder analysis can also help with this KSA.

You can make the identification of ethical issues as challenging or simple as you want – or, use what you know about Bloom's taxonomy to help yourself improve your analyses.

While a very complex thing to do, the simplest way to use GLs and CE is to match the words from the GL/CE to the problem you're considering. If you can't match exactly, you can use synonyms: the matching is lower in cognitive complexity, but an excellent way to get familiar with the GLs/CE and also to begin to explore what ethical practice is actually all about.

Next most complex way to use the GLs and CE is to consider both the letter of each element (exact words) and also the <u>intent</u>. The GLs and CE have to be general – both ASA and ACM seek to generally describe the ethical practitioner. It would be impossible to describe more specifically all of the things the ethical practitioner does, or how they do each task at work in an ethical way. But this necessary generality means that the reader needs to bring some cognitive power to their use of these documents to guide what they do and how they do it! If you recall, the difference between the higher/more complex Bloom's level behaviors (levels 4-6) and the lower ones is a decrease in the use/utility of rules and the fact that levels 5 and 6 require judgment, because rules may not suffice or exist. All of ethical practice requires these very complex bloom's level functions – so, to prepare yourself to get all the way there, you can consider both the explicit **and the implicit** features of the CE/GL items to guide your decision making, reasoning, and practice.

The most complex way to use the GLs and CE is, of course, to use them as they were intended – guidance. Leaving Bloom's 4 (application of rules/prediction) behind and focusing on the intent (using Bloom's 5-6 and evaluating, and synthesizing diverse sources of information with your experience) might make it seem that every single element or principle could potentially be interpreted as applying to all tasks – and on the one hand, that's an excellent place for the practitioner to be! If all statistics and data science practitioners sought to follow the intent of the GL/CE in all tasks, it would make the data-centered world a much better place. On the other hand, considering the real intent behind the GLs/CE should lead the practitioner to recognize that the reason for multiple elements in each document is because it isn't really helpful to have every single element (72 for ASA, 25 for ACM) be applied in every single task. Not only is it very complex to use Bloom's 5-6 at all, it is particularly challenging to use 5-6 to help you focus on the most appropriate features of the GL/CE for a given task. Note that, for the same task in different contexts/circumstances, the "most appropriate feature of the CE/GL" can vary. That is part of the judgment you bring to your work, and also part of why this is the most complex way to use the GLs/CE.

It is important to note that identifying or recognizing an ethical issue may happen as you prepare to do your work, before anything has "gone wrong". By identifying problems and addressing them, which is the DSEC perspective and is also alluded to throughout the ACM CE, the recognition or identification of an ethical issue may far more frequently arise when you must decide what to do in response to a specific order, request, or situation that was not created by you. The following four tables describe how ethical challenges can be identified by taking a single case (example behavior from a real work experience) and exploring it in the context of the ACM and ASA practice standards to determine whether the example behavior is recognizable as inconsistent with either of them.

Case 1A. The client/collaborator does not know how to present, or is not committed to presenting, the correct, transparent interpretation of results.

ASA GL guidance

ASA Guideline Principle	Principle-identified challenges/decisions to be made
A. Professional integrity and accountability B. Integrity of data and methods	The data analyst will execute their professional obligations to the best of their abilities; no ethical challenges are identified using either of these Guideline Principles.
C. Responsibilities to Stakeholders	*Challenge*: the analyst has obligations – to science and to the public (and to a funder if funding is involved) to ensure that the collaborator/client uses (interprets, presents) their results responsibly.
D. Responsibilities to research subjects, data subjects, or those directly affected by statistical practices, Data Subjects, or those directly affected by statistical practices	Because the data were already collected, as the analyst fulfills Guideline Principles A & B, Responsibilities to research subjects, data subjects, or those directly affected by statistical practices are met.
E. Responsibilities to members of multidisciplinary teams	*Challenge*: the analyst has obligations to all members of the research team to ensure that the collaborator/client uses (interprets, presents) their results responsibly.
F. Responsibilities to fellow statistical practitioners and the profession	*Potential solution:* This Guideline Principle can be used to help encourage collaborator/client responsible use of the statistical results.
G. Responsibilities of Leaders, Supervisors, and Mentors in Statistical Practice	*Potential solution:* This Guideline Principle can be used to help strengthen the resolve of the collaborator/client to use the statistical results *responsibly*.
H. Responsibilities regarding potential misconduct	*Challenge*: the analyst's obligations to ensure that the collaborator/client uses their results responsibly implies that, if misconduct is encountered, the analyst has additional obligations that will arise.

Table 1.9.1.[19] *Using the ASA Ethical Guidelines for Professional Practice to identify ethical challenge(s). Case 1A: ASA*

---

[19] Table is adapted from Gunaratna NS & Tractenberg RE. (2016). Ethical reasoning with the 2016 revised ASA Ethical Guidelines for Statistical Practice. Proceedings of the 2016 Joint Statistical Meetings, Chicago, IL. Pp. 3763-3787.

Just as the ASA GLs enable the identification of challenges to ethical practice and help to find solutions, the ACM CE can also be utilized, in the same case, to identify potential ethical challenges and solutions.

Case 1B. The client/collaborator does not support/is not committed to the correct, transparent reporting and documentation of the system, model, or algorithm under development.

ACM CE Guidance

ACM CE Principle	Principle-identified challenges/decisions to be made
1. General Ethical Principles	*Challenge*: The ethical computing professional encourages transparency and accountability – which are impossible when correct and transparent documentation are prevented. The ability to determine whether or not a system causes harm is essential, and preventing transparent documentation undermines that. "A computing professional has an additional obligation to report any signs of system risks that might result in harm." *Potential solutions*: "If leaders do not act to curtail or mitigate such risks, it may be necessary to "blow the whistle" to reduce potential harm."
2. Professional Responsibilities	*Challenge*: Impediments to transparent documentation of a system undermine the ethical practitioner's obligations: "Computing professionals should respect the right of those involved to transparent communication about the project. Professionals should be cognizant of any serious negative consequences affecting any stakeholder that may result from poor quality work and should resist inducements to neglect this responsibility." *Potential solution:* "In cases where misuse or harm are predictable or unavoidable, the best option may be to not implement the system."
3. Professional Leadership Principles	*Potential solution:* "Leaders should pursue clearly defined organizational policies that are consistent with the Code and effectively communicate them to relevant stakeholders." It may be the case that leaders have not

	communicated how this action is inconsistent with the CE. Engagement with leadership to formulate a response whereby the client is convinced to stop preventing practitioners from following the CE is suggested by Principle 3.
4. Compliance with the Code	*Challenge:* Principle 4 describes how failures to comply with the CE are the responsibility of all ethical computing professionals, because they have an obligation to "Uphold, promote, and respect the principles of the Code." (ACM 4.1). While the client may not meet even the minimum test the ACM CE provides for whose behavior they should guide ("The Code is designed to inspire and guide the ethical conduct of all computing professionals, including current and aspiring practitioners, instructors, students, influencers, and *anyone who uses computing technology in an impactful way*", emphasis added), Principle 4 charges those whose behavior the CE clearly should guide with taking action to resolve ethical issues (like this one) and to encourage adherence to the CE by all practitioners. Three ethical challenges are clearly identifiable according to Principle 4: permitting the client to prevent the practitioner from adhering to the CE; not encouraging the client to adhere to the CE themselves; and not encouraging leaders to support efforts to permit full and transparent documentation of the system. These additional, compliance-based ethical issues arise because failure to fully and transparently document a system represents an ethical lapse on the computing professional's part.

Table 1.9.2. *Using the ACM Ethical Guidelines for Professional Practice to **identify ethical challenge(s)**. Case 1B: ACM*

A second pair of examples features a different case with examples of how the ASA and/or ACM can be utilized to both identify ethical challenges that the case represents and also solutions.

Case 2A. Client/collaborator is not aware of statistical concepts of sampling, bias, etc. and may not be able to identify important confounders or sources of bias. Their data are already collected/data collection mechanisms are already deployed, and you are committed to planning and executing the analysis.

ASA GL guidance

ASA Guideline Principle	Principle-identified challenges/decisions to be made
A. Professional integrity and accountability	*Challenge*: The ethical data analyst will execute their professional obligations to the best of their abilities; if the data were collected in a biased way, analyses and the interpretations that are supportable are very limited.
B. Integrity of data and methods	*Challenge*: The ethical data analyst will analyze the data in the manner that is most consistent with the integrity of the data – in order to minimize bias and maximize transparency. This is particularly important when the analyses will be used to support policy decisions.
C. Responsibilities to Stakeholders	*Potential solution*: This Guideline Principle can be used to help encourage collaborator/client to accept the most appropriate analyses/results given the data – unifying the team's commitment to the responsible use of the statistical results.
D. Responsibilities to research subjects, data subjects, or those directly affected by statistical practices, Data Subjects, or those directly affected by statistical practices	Because the data were already collected, as the analyst fulfills Guideline Principles A & B, Responsibilities to research subjects, data subjects, or those directly affected by statistical practices are met.
E. Responsibilities to members of multidisciplinary teams	*Challenge*: If there is a conflict, the analyst must prioritize their obligations to the profession (F) and to stakeholders (C) over other perceived obligations to members of the research team by analyzing the data using methods that are appropriate given the data and their origins.

F. Responsibilities to fellow statistical practitioners and the profession	*Potential solution:* This Guideline Principle can be used to help encourage collaborator/client responsible use of the statistical results.
G. Responsibilities of Leaders, Supervisors, and Mentors in Statistical Practice	*Potential solution:* the analyst's obligation to avoid misconduct, and to avoid condoning or appearing to condone statistical, scientific, or professional misconduct (G.1) can be used to help explain the critical nature of transparency and defensible interpretations of statistical results –particularly as they derive from potentially biased data.
H. Responsibilities regarding potential misconduct	*Potential solution:* This Guideline Principle can be used to help engage a collaborator/client in discussions about the documented limitations on the statistical results that derive from the data collection procedures -to protect against charges of misconduct or detrimental research practices.

Table 1.9.3. *Using the ASA Ethical Guidelines for Professional Practice to **identify** ethical challenge(s).* *Case 2A: ASA*

As with the first example case, this case can also be explored with the ACM CE.

Case 2B. Client/employer has not identified important confounders or sources of bias in the data they are collecting. Their data collection mechanisms are already deployed, and you are committed to creating a system to execute the analysis of the data.

ACM CE Guidance

ACM CE Principle	Principle-identified challenges/decisions to be made
1. General Ethical Principles	*Challenge:* The ethical computing has an obligation to avoid harm, and "should follow generally accepted best practices unless there is a compelling ethical reason to do otherwise. Additionally, the consequences of data aggregation and emergent properties of systems should be carefully analyzed." Since harms may accrue when bias is not recognized, the data being collected has a real likelihood of resulting in harms. Moreover, "computing professionals

should take action to avoid creating systems or technologies that disenfranchise or oppress people. Failure to design for inclusiveness and accessibility may constitute unfair discrimination."

Potential solutions: "Computing professionals should only use personal information for legitimate ends and without violating the rights of individuals and groups. This requires taking precautions to prevent re- identification of anonymized data or unauthorized data collection, ensuring the accuracy of data, understanding the provenance of the data, and protecting it from unauthorized access and accidental disclosure. Computing professionals should establish transparent policies and procedures that allow individuals to understand what data is being collected and how it is being used, to give informed consent for automatic data collection, and to review, obtain, correct inaccuracies in, and delete their personal data." and, if necessary, "If leaders do not act to curtail or mitigate such risks, it may be necessary to "blow the whistle" to reduce potential harm."

2. Professional Responsibilities

Challenge: Knowingly designing a system that may result in harms or bias (which then may lead to harms) violates Principle 2: "Computing professionals should respect the right of those involved to transparent communication about the project. Professionals should be cognizant of any serious negative consequences affecting any stakeholder that may result from poor quality work and should resist inducements to neglect this responsibility."

Potential solution: "In cases where misuse or harm are predictable or unavoidable, the best option may be to not implement the system." The computing professional must balance this potential solution against ACM 1.3: "commitments should be honored." The ethical computing professional does not honour commitments (i.e., to create the system in this case) if it becomes clear that they will lead to violations of other ACM CE principles – particularly when these lead to harms.

3. Professional Leadership Principles	*Potential solution*: "Leaders should pursue clearly defined organizational policies that are consistent with the Code and effectively communicate them to relevant stakeholders." It may be the case that leaders have not communicated how this system for analysis would be inconsistent with the CE. Engagement with leadership to formulate a response whereby the client is convinced to reconsider how their data are being collected is suggested by Principle 3. Even if this will delay the results and/or cost more than planned, the mitigation of harms and the real risks of harms due to unknown biases in the data is a principal consideration for the ethical computing professional, and leaders are obliged to promote compliance with the CEs.
4. Compliance with the Code	*Challenge:* Principle 4 describes how failures to comply with the CE are the responsibility of all ethical computing professionals, because they have an obligation to "Uphold, promote, and respect the principles of the Code." (ACM 4.1). While the client may not meet even the minimum test the ACM CE provides for whose behavior they should guide ("The Code is designed to inspire and guide the ethical conduct of all computing professionals, including current and aspiring practitioners, instructors, students, influencers, and *anyone who uses computing technology in an impactful way*", emphasis added), Principle 4 charges those whose behavior the CE clearly should guide with taking action to resolve ethical issues (like this one) and to encourage adherence to the CE by all practitioners. Not encouraging the client to adhere to the CE, and/or not encouraging leaders to support efforts to permit full and transparent documentation of the system, both represent an ethical lapse on the computing professional's part.

Table 1.9.4. *Using the ACM Ethical Guidelines for Professional Practice to **identify ethical challenge(s)**. Case 2B: ACM*

These cases are very brief, but are authentic, as they derive from actual professional experience. As you can see, examination of the GL and CE can support KSAs 3 (identifying the ethical issue) and 4 (formulating alternative responses to that ethical challenge). We apply Stakeholder Analysis in Section 2 to familiarize ourselves with all of the GL principles, considering how these principles and elements can be brought to bear to ensure that how we do practice (and think about our practice/engagement with data) are ethical. In Section 3 we explore ethical reasoning when there actually IS the potential for an ethical issue to have arisen in the vignette.

Chapter 1.10
Identify alternative actions
(on the ethical issue): KSA 4

Once the prerequisite knowledge has been accumulated, whether or not an ethical challenge is identified should be clear. While identifying the ethical issue is possibly <u>the</u> hardest ER KSA, only slightly more challenging will be determining what are plausible alternatives you can take to address, or respond to, whatever ethical challenge you identified. One reason why this particular KSA can be so problematic for the computing or statistical practitioner is that "do nothing" or "act like nothing is wrong", or "ignore it/hope it will go away/hope someone with more authority recognizes the problem and makes a decision" actually represent responses to an ethical dilemma - even though, counterintuitively, they involve "no action" on the part of the computing or statistical practitioner. Once you recognize and accept that "do nothing" is an actual decision – one that is made repeatedly, daily, all around the world by people faced with ethical dilemmas that they themselves did not create, or that they are worried about acknowledging, it is simpler to perceive other, alternative, responses.

There are literally **always** two decisions that can be made in any circumstance: do *nothing* or "do something". However, it can be difficult to figure out which of these is better – i.e., more consistent with our decision-making framework – and in the case of "do something", what exactly TO do is not clear.

Exploring these two options further, particularly "do something", can lead to three possible decisions that can be made in *any* circumstance:

a) do nothing.
b) consult or confer with a peer or a supervisor – using the GL or CE, or other resources (policies in your organization, for example).
c) report violations of policy, procedure, ethical guidelines, or law.

"In cases where a practitioner does not feel sufficiently empowered or otherwise able to discuss an ethical challenge directly with a client, engaging a colleague in formulating a response is a viable alternative. The chosen colleague could be one with greater experience or sophistication in reasoning through or responding to an ethical challenge, or could have a different relationship with the client (e.g., the colleague could be a professor who serves as a mentor to a student practitioner)." (Gunaratna & Tractenberg 2016, p. 3775). As we have seen, the ACM CE specify that ongoing conversations about ethical practice are to be supported – so any time you choose option b (consult or confer), that alternative is likely to be at least partly conforming to the ACM CE. You will also recall that the ACM specifically set aside two obligations of the computing professional to at least consider reporting violations of the CE to the individual who you think is violating the CE (ACM 4.1) and to the ACM itself (ACM 4.2). So, both of options b and c are consistent with the CE. As we will see in the next examples, "do nothing" is a response to an ethical issue that you observe, or perceive may occur, and is inconsistent with both the ACM CE and ASA GLs.

The following four Tables outline using the ASA Ethical Guidelines for Statistical Practice (Cases 1A, 2A) or ACM Code of Ethics (Cases 1B, 2B) to *articulate and evaluate alternative actions*.

Case 1A. The client/collaborator does not know how to present, or is not committed to presenting, the correct, transparent interpretation of results.

Alternative actions:

Do nothing.
Engage a colleague to formulate a response.
Report the client/collaborator.

ASA Guideline Principle	Principle-identified alternative actions and their evaluation
A. Professional integrity and accountability B. Integrity of data and methods	**Do nothing**: this alternative is *not consistent with Principle A.* **Engage a colleague** to formulate a response: consistent with Principle A **Report the client/collaborator**: Principle A is not informative about this alternative. Principle B is *not informative* for any of these alternatives.
C. Responsibilities to Stakeholders	**Do nothing**: this alternative is *not consistent with Principle C.* **Engage a colleague** to formulate a response: consistent with Principle C. **Report the client/collaborator**: consistent with Principle C, but only with simultaneous application of Principle H as well.
D. Responsibilities to research subjects, data subjects, or those directly affected by statistical practices, Data Subjects, or those directly affected by statistical practices	Principle D is not particularly informative for any of these alternatives; however, **doing nothing** to address an ethical challenge tends to prioritize other stakeholders over the research subjects –implicitly or explicitly.
E. Responsibilities to members of multidisciplinary teams	**Do nothing**: this alternative is *not consistent with Principle E.* **Engage a colleague** to formulate a response: consistent with Principle E, one option is to engage others on the research team to formulate a response. **Report the client/collaborator**: consistent with Principle E, but only with simultaneous application of Principle H as well and possible, consideration of the ethical obligations of these other team members.

F. Responsibilities to fellow statistical practitioners and the profession	**Do nothing**: this alternative is *not consistent with Principle F.* **Engage a colleague** to formulate a response: consistent with Principle F and other statisticians may have similar experiences to share. **Report the client/collaborator**: Principle F is not specifically informative about this alternative, but the analyst in the situation may be able to prevent other analysts from being entangled in a similar situation by formally reporting the situation.
G. Responsibilities of Leaders, Supervisors, and Mentors in Statistical Practice	**Do nothing**: Principle G is not informative about this alternative action. **Engage a colleague** to formulate a response: Principle G may be informative about this alternative. **Report the client/collaborator**: Principle G is not specifically informative about this alternative action.
H. Responsibilities regarding potential misconduct	**Do nothing**: this alternative is *not consistent with Principle H.* **Engage a colleague** to formulate a response: consistent with Principle H, although everyone who is consulted must also understand their responsibilities regarding what actually constitutes misconduct. **Report the client/collaborator**: consistent with Principle H.

Table 1.10.1 [20] . *Using the ASA Ethical Guidelines for Statistical Practice to* *articulate and evaluate alternative actions. Case 1A: ASA*

---

[20] Table is adapted from Gunaratna NS & Tractenberg RE. (2016). Ethical reasoning with the 2016 revised ASA Ethical Guidelines for Statistical Practice. Proceedings of the 2016 Joint Statistical Meetings, Chicago, IL. Pp. 3763-3787.

Case 1B. The client/collaborator does not support/is not committed to the correct, transparent reporting and documentation of the system, model, or algorithm under development.

Alternative actions:

Do nothing/allow client's perspective and do not report.
Engage a colleague to formulate a response.
Refuse to continue on the project.

ACM CE Principle	Principle-identified alternative actions and their evaluation
1. General Ethical Principles	**Do nothing**: this alternative is *contrary to* ACM Principle 1. **Engage a colleague** to formulate a response: consistent with Principle 1 as well as the preamble (open and transparent conversations) **Refuse to continue on the project**: consistent with Principle 1, but CE suggest notification – i.e., either reporting to the ACM or to the client (or both) that the behavior is contrary to ACM CE and that is why you will no longer participate.
2. Professional Responsibilities	**Do nothing**: this alternative is *contrary* to ACM Principle 2. **Engage a colleague** to formulate a response: consistent with Principle 2 as well as Principle 1 and the preamble (open and transparent conversations) **Refuse to continue on the project**: consistent with Principle 2, but CE suggest documentation of the harm/potential for harm this behavior represents, and that this is contrary to ACM CE and that is why you will no longer participate.

3. Professional Leadership Principles	**Do nothing**: this alternative is *contrary* to ACM Principle 3: specifically, "Leaders should pursue clearly defined organizational policies that are consistent with the Code and effectively communicate them to relevant stakeholders."
	Engage a colleague to formulate a response: consistent with Principle 3 and may be the most ACM CE consistent alternative, particularly if leadership is involved to help resolve the issue (supporting the full and transparent documentation of the system).
	Refuse to continue on the project: consistent with Principle 3, but CE suggest that the leadership decision to stop participating should be or become part of policies and processes that specifically reflect the CE.
4. Compliance with the Code	**Do nothing**: this alternative violates ACM Principle 4: specifically, "Computing professionals who recognize breaches of the Code should take actions to resolve the ethical issues they recognize, including, when reasonable, expressing their concern to the person or persons thought to be violating the Code." Doing nothing not only appears to encourage the violation of the CE (failing to fully document a system, failing to make it transparent and evaluable) but also represents a failure to take action as ACM 4.1 states should be done.
	Engage a colleague to formulate a response: consistent with Principle 4, representing taking "actions to resolve the ethical issues they recognize" (ACM 4.1).
	Refuse to continue on the project: Unless it is accompanied by documentation of why, refusing to continue is actually a violation of ACM 4. Not encouraging the client to adhere, and/or not encouraging leaders to support efforts to permit full and transparent documentation of the system, represents an ethical lapse on the computing professional's part, compounding what was originally only a violation of the ACM CE by the client.

Table 1.10.2. *Using the ACM Code of Ethics to **articulate and evaluate alternative actions**. Case 1B: ACM*

Case 2A. Client/collaborator is not aware of statistical concepts of sampling, bias, etc. and may not be able to identify important confounders. Their data are already collected, and you are committed to planning and executing the analysis.

Alternative actions:

Do nothing.
Engage a colleague to formulate a response.
Report the client/collaborator.

ASA Guideline Principle	Principle-identified alternative actions *and their evaluation*
A. Professional integrity and accountability	**Do nothing**: this alternative is *directly opposed to* Principle A; professional integrity is required to plan and execute a defensible, appropriate analysis. **Engage a colleague** to formulate a response: consistent with Principle A. **Report the client/collaborator**: Principle A is not informative about this alternative.
B. Integrity of data and methods	**Do nothing**: this alternative is *directly opposed to Principle B; combined with Principle A,* the analyst is obliged to plan and execute an analysis consistent with the data. **Engage a colleague** to formulate a response: consistent with Principle B. **Report the client/collaborator**: Principle B is not informative about this alternative.
C. Responsibilities to Stakeholders	**Do nothing**: this alternative is *not consistent with Principle C.* **Engage a colleague** to formulate a response: consistent with Principle C. **Report the client/collaborator**: consistent with Principle C, but only with simultaneous application of Principle H as well.
D. Responsibilities to research subjects, data subjects, or those directly affected by statistical practices, Data Subjects, or those directly affected by statistical practices	Principle D is not particularly informative for any of these alternatives; however, **doing nothing** to address an ethical challenge tends to prioritize other stakeholders over the research subjects – whether implicitly or explicitly.

E. Responsibilities to members of multidisciplinary teams	**Do nothing**: this alternative is *not consistent with Principle E.* **Engage a colleague** to formulate a response: consistent with Principle E, one option is to engage others on the research team to formulate a response. **Report the client/collaborator**: consistent with Principle E, but only with simultaneous application of Principle H as well and possible, consideration of the ethical obligations of these other team members.
F. Responsibilities to fellow statistical practitioners and the profession	**Do nothing**: this alternative is *not consistent with Principle F.* **Engage a colleague** to formulate a response: consistent with Principle F and other statisticians may have similar experiences to share. **Report the client/collaborator**: Principle F is not specifically informative about this alternative, but the analyst in the situation may be able to prevent other analysts from being entangled in a similar situation by formally reporting the situation.
G. Responsibilities of Leaders, Supervisors, and Mentors in Statistical Practice	**Do nothing**: this alternative is *not consistent with Principle G.* The analyst can, at a minimum, inform the client/collaborator of their responsibilities when data are insufficiently documented (as in this case). **Engage a colleague** to formulate a response: Principle G may be informative about this alternative. **Report the client/collaborator**: Principle G is not specifically informative about this alternative action.
H. Responsibilities regarding potential misconduct	**Do nothing**: this alternative is *not consistent with Principle H, but* that is not because the client/collaborator is engaging in misconduct necessarily. Principle H is *always* relevant. **Engage a colleague** to formulate a response: consistent with Principle H, although everyone who is consulted must also understand their responsibilities regarding what actually constitutes misconduct. **Report the client/collaborator**: consistent with Principle H.

Table 1.10.3. *Using the ASA Ethical Guidelines to* **articulate and evaluate alternative actions.** *Case 2A: ASA*

Case 2B. Client/employer has not identified important confounders or sources of bias in the data they are collecting. Their data collection mechanisms are already deployed, and you are committed to creating a system to execute the analysis of the data.

Alternative actions:

Do nothing/allow client's perspective and do not report.
Engage a colleague to formulate a response.
Refuse to continue on the project.

ACM CE Principle	Principle-identified alternative actions and their evaluation
1. General Ethical Principles	**Do nothing**: this alternative is *contrary to* ACM Principle 1.
	Engage a colleague to formulate a response: consistent with Principle 1 as well as the preamble (open and transparent conversations). Note that the response may be in the form of policy, "whistle-blowing", or withdrawal from the project. Engaging in an open discussion about the ethical implications of this situation can strengthen the profession *and also* yield a solution.
	Refuse to continue on the project: mostly consistent with Principle 1, but CE suggest that it is not sufficient to simply withdraw or refrain from a project/system where bias and/or harms can be predicted; specifically, "If leaders do not act to curtail or mitigate such risks, it may be necessary to "blow the whistle" to reduce potential harm." That is, simply withdrawing is not a sufficient action when behavior is contrary to ACM CE; notifying the client – and possibly others – as to why you will no longer participate is an essential aspect of this alternative.
2. Professional Responsibilities	**Do nothing**: this alternative is *contrary to* ACM Principle 2.
	Engage a colleague to formulate a response: consistent with Principle 2. "In cases where misuse or harm are predictable or unavoidable, the best option may be to not implement the system." The computing professional must balance this potential solution against ACM 1.3: "commitments should be honored." Engaging in an open

discussion about the ethical implications of this situation can lead to a solution that balances these two obligations.

Refuse to continue on the project: mostly consistent with Principle 2, "Professionals should be cognizant of any serious negative consequences affecting any stakeholder that may result from poor quality work and should resist inducements to neglect this responsibility." That is, simply withdrawing is not a sufficient action when behavior is contrary to ACM CE; the behavior should (also) be resisted. Notifying the client – and possibly others – as to why you will no longer participate is an essential aspect of this alternative.

| 3. Professional Leadership Principles | **Do nothing**: this alternative is *contrary to* ACM Principle 3.
Engage a colleague to formulate a response: *mostly* consistent with Principle 3. Leaders have a specific obligation to ensure that their organizational policies are clear, consistent with the CE, and are communicated to all relevant stakeholders. In this alternative, then, "engage a colleague" actually refers to the leader ensuring that all colleagues, peers, and employees understand the CE and how the organization and its policies are all consistent with the CE. With that caveat, this may be the alternative that addresses ACM CE 3 best.
Refuse to continue on the project: partially consistent with Principle 3. Simply withdrawing is not a sufficient action when client behavior is contrary to ACM CE; the behavior should be actively resisted, and policies introduced, strengthened, or reiterated to all members of the organization and other stakeholders, notifying the client – and others throughout the organization– as to why this client's behavior cannot be permitted to drive the organization or any of its employees/members to violate the CE. |
| 4. Compliance with the Code | **Do nothing**: this alternative is *contrary to* ACM Principle 4. The client is violating the CE and seeks to cause computing professionals to |

violate the CE as well. Doing nothing compounds the client's violation and adds to the computing professional's.

Engage a colleague to formulate a response: consistent with Principle 4, as long as the response is consistent with the CE (because ACM 4 is all about following the CE).

Refuse to continue on the project: partially consistent with Principle 4. Simply withdrawing is not a sufficient action when client behavior is contrary to ACM CE; the behavior should be actively resisted, and the client informed of their actions' violation of the CE. Failing to discourage the client from continuing to violate –and induce others to violate –the ACM CE compounds the client's violation, and adds to the computing professional's.

Table 1.10.4. *Using the ACM Code of Ethics to **articulate and evaluate alternative actions**. Case 2B: ACM*

As can be seen in each of these examples, "doing nothing" when faced with an ethical challenge **is an alternative action in every case** – and this specific alternative is *inconsistent* with every ASA Ethical Guideline Principle and every ACM Code of Ethics Principle as well. However, it is essential to recognize that "doing nothing" when faced with an ethical challenge is an actual decision – and this decision is contrary to the ASA and ACM practice standards. You can also see from these examples that engaging a colleague is, in some cases, the alternative that is most consistent with the GLs and CE. Of course, the response that is formulated when you engage colleagues cannot be "do nothing"; and you can also see that the ACM CE tends to support engagement with colleagues to ensure that both the client and others in the organization are made aware that the specific client behaviors are unacceptable violations of the CE.

The two admittedly generic alternatives ("engage a colleague to formulate a response" and "report the client/collaborator" (ASA) or "withdraw" (ACM)) are *evaluable* – although when an individual is actually engaging in the steps of ethical reasoning to address a challenge they encounter during a collaboration, more specific alternative actions will very likely be identified. Note also that simply withdrawing from the project might be a viable alternative (which we did not consider) when completing this KSA using the ASA GLs, but because of the specific obligations to protect the rights of stakeholders, and to ensure compliance with the ACM CE, simply withdrawing is never an acceptable option for those following the ACM CE: in each case, "withdrawal" is always to be accompanied with some documentation of why the client's behavior is a) inconsistent with the CEs and b) causing the withdrawal. This is a particular feature of ACM CE 4 (Compliance with the Code).

Also, it should be pointed out that the ACM CE discusses "whistleblowing" in several places, where leadership, the public, other stakeholders, or the ACM (or some combination of these) are to be notified of the potential (or actual) harms that a computing system may cause. This extends to behaviors that are contrary to the CE or more specifically, behaviors that represent some stakeholder or other entity trying to get a computing professional to violate the CE. However, in any work environment there will be policies where any ethical or policy violations, irrespective of the ACM CE/ASA GL, must be reported. There are federal protections offered for whistleblowers, and whenever you notice/recognize that a *law* is being/has been violated, "reporting" is clearly required! For ethical issues that arise when anyone violates procedures or guidelines that the organization has established, reporting may also be required. At a minimum, if you find yourself working in an environment where you feel pressured to violate any law or policy, or the ASA GL/ACM CE, it is well worth considering whether *or not* that is an environment that will enable you to work in an ethical capacity.

Chapter 1.11
Make and justify decision,
and reflect on that decision: (KSAs 5-6)

At this point, we have pretty fully explored KSAs 1 and 2, and spent time on KSAs 3 and 4. While we focused on KSA 4, and specifically, how "do nothing" is never an alternative that the practice standards will recommend or support when there are real harms that may accrue as a result, it is difficult to really conceptualize how all the KSAs work together without an actual case. We will continue to defer that until Section 3.

Here we focus on re-visiting all the results from the first two KSAs: how your prerequisite knowledge (practice standards and SHA) informs the decision-making framework, and these two KSAs are interrelated in their support for how the eventual decision will be made (KSA 5), particularly because part of "making" the decision is ensuring that you can justify that decision. Our emphasis on KSAs 1 and 2 will pay off there, because the justification for your decision will come from the prerequisite knowledge you pulled together, and will be bolstered by your selection of the decision-making framework that features that prerequisite knowledge.

The decision that you make will obviously be one of the alternatives you identified in KSA 4. The justification of that decision will involve your discussing the evidence/harms or benefits calculus that informed your decision, and the ASA GL/ACM CE considerations, and stakeholder effects. That is, making and justifying the decision will utilize all of your prerequisite knowledge, summarized with either the Utilitarian or the Virtue framework – or both, especially if they support different decisions, and you are selecting both an alternative *and* the perspective that fits the problem best). The key dependence of the decision (and its justification) on the work you did in KSAs 1-2 is part of why we spent so much time on those KSAs! Articulating the decision –what to do in the face of the ethical challenge that was identified – must include at least some discussion of how stakeholder effects were considered. Stakeholders and benefits/harms must be considered as a result of the decision and can help to arrive at (a better) one, where "better" would be "resulting in fewer or less egregious harms, or in more or more substantial benefits, or both".

You can see from the examples in the previous two chapters that, while KSAs 3 and 4 can be difficult, once those have been carried out, you make your selection from the alternative actions (KSA 4) and then you justify this decision based on your summary and analysis of all the information you compiled in KSAs 1-3.

Based on the tables in the previous chapter, we can summarize KSAs 1-2 in our contemplation of what the ethical problem actually is (KSA 3). It is essential to be specific about what the problem is so you can articulate and then contemplate or evaluate the alternatives available to you (KSA 4). Once you have alternatives worked out, making and justifying the decision (KSA 5) is straightforward. Documenting your decision is important because if new evidence comes to light, or if a similar situation occurs in the future, or even if a new ethical practice standard or policy becomes available, the decision can be revisited -and updated in a reproducible, defensible, way.

Case 1A, ASA Ethical Guidelines. *The client/collaborator does not know how to present, or is not committed to presenting, the correct, transparent interpretation of results.*

KSA 3: Identifying the ethical problem (according to the ASA GLs)

Using the ASA GLs, we can identify ethical problems with this case just by identifying which ASA elements this behavior seems to conflict with, contradict, or specifically prevent the statistical practitioner from doing/following.

According to the ASA GLs:

A. Professional integrity and accountability: indirectly, this action threatens to violate A2, (the ethical statistical practitioner) "Uses methodology and data that are valid, relevant, and appropriate, without favoritism or prejudice, and in a manner intended to produce valid, interpretable, and reproducible results" and A4: "Opposes efforts to predetermine or influence the results of statistical practices, and resists pressure to selectively interpret data" Ethical statistical practice requires the use of relevant methods, no matter what the client/collaborator wishes; and the correct and complete report/presentation of valid interpretations of the results. This client may seek to influence "the results" by preventing others from knowing what they are, or the extent of their limitations. Or the client may ask that different methods be used (tried) until a satisfactory answer (for them) is obtained.

B. Integrity of data and methods: B2: "Is transparent about assumptions made in the execution and interpretation of statistical practices including methods used, limitations, possible sources of error, and algorithmic biases. Conveys results or applications of statistical practices in ways that are honest and meaningful." And B3 "Communicates the stated purpose and the intended use of statistical practices. Is transparent regarding a priori versus post hoc objectives and planned versus unplanned statistical practices. Discloses when multiple comparisons are conducted, and any relevant adjustments." This communication (B2/B3) should be reported to the client, but if the client intends to publish or otherwise present results, the statistician's commitment to valid interpretations and transparency cannot be impeded. The transparency and communication responsibilities extend beyond the client to those for whom the results are intended.

C. Responsibilities to Stakeholders: Challenge: the analyst has obligations to this client/collaborator, but they are not the only stakeholder. Other important stakeholders include science and the public, who make up parts of "those who fund, contribute to, use, or are affected by statistical practices are considered stakeholders". The responsibilities are met by ensuring that the collaborator/client uses (interprets, presents) their results responsibly: Regardless of personal or institutional interests or external pressures, does not use statistical practices to mislead any stakeholder" (C2). The client or collaborator who does not support valid inferences and transparency also does not support ethical statistical practice in general, but the ethical statistical practitioner cannot be prevented from fulfilling their obligation to stakeholders by the decisions or inclinations of the client. C8 specifies, "prioritizes both scientific integrity and the principles outlined in these Guidelines when interests are in conflict."

D. Responsibilities to research subjects, data subjects, or those directly affected by statistical practices, Data Subjects, or those directly affected by statistical practices: the primary way the client's unwillingness to be transparent affects the ethical statistical practitioner is by potentially impeding the responsibility to follow this principle: "The ethical statistical practitioner does not misuse or condone the misuse of data. They protect and respect the rights and interests of human and animal subjects. These responsibilities extend to those who will be directly affected by statistical practices." Incomplete or incorrect reporting of results constitutes a misuse of data, so would cause an abrogation of the responsibilities in D.

E. Responsibilities to members of multidisciplinary teams: the analyst has obligations to all members of the research team to ensure that the collaborator/client uses (interprets, presents) their results responsibly: "E2. Ensures all discussion and reporting of statistical design and analysis is consistent with these guidelines." And "E3. Avoids compromising scientific validity for expediency." And "E4. Strives to promote transparency in design, execution, and reporting or presenting of all analyses."

F. Responsibilities to fellow statistical practitioners and the profession: "F1 Makes documentation suitable for replicate analyses, metadata studies, and other research by qualified investigators." And F3: "Instills in students and non-statisticians an appreciation for the practical value of the concepts and methods they are learning or using."

G. Responsibilities of leaders, supervisors, and mentors in statistical practice: G1 states that those who lead, supervise, or mentor have a special obligation to "Ensure appropriate statistical practice that is consistent with these Guidelines. Protect the statistical practitioners who comply with these Guidelines, and advocate for a culture that supports ethical statistical practice." In some cases, the statistical practitioner working with a client or collaborator is – or can be considered to be -leading the data analysis/data science part of the project. Advocacy for a culture that supports ethical statistical practice includes standing up for ethical behavior- even when this is requested or required by the client or collaborator.

H. Responsibilities regarding potential misconduct: The ethical statistical practitioner has obligations to ensure that the collaborator/client uses their results responsibly. Specifically, H2 states that the ethical statistical practitioner "Avoids condoning or appearing to condone statistical, scientific, or professional misconduct. Encourages other practitioners to avoid misconduct or the appearance of misconduct." This specific item underscores how it is never ethical to ignore client or collaborator misconduct, or to "do nothing" when faced with misconduct and detrimental research practices.

Summary of KSAs 1-3: The client/collaborator in this vignette is violating, or would violate, several of the ASA GL principles and guidelines, and may also impede (or seek to impede) the ethical statistical practitioner from fulfilling their GL-specified responsibilities. While the client/collaborator may in fact be acting consistently with their own professional practice standards (and the ethical statistical practitioner recognizes this, per E1) failures to transparently report all results by collaborators cause the statistician to violate E2, G1 and many other Principles and elements as outlined above. Even if the client is following their profession's "norms", failures of transparency or actually engaging in statistical misconduct (i.e., failures of transparency and responsibilities of the statistical practitioner per Principles A, B, E and F) fall into the *detrimental research practices* category outlined in NASEM 2017 (especially Section 2 "Research Misconduct and Detrimental Research Practices" and Ch 7). Such practices are discussed in ASA GL Principle H, particularly H2 (the ethical statistical practitioner "Avoids condoning or appearing to condone statistical, scientific, or professional misconduct. Encourages other practitioners to avoid misconduct or the appearance of misconduct"), and Principle G (G1: "Ensure appropriate statistical practice that is consistent with these Guidelines. Protect the statistical practitioners who comply with these Guidelines, and advocate for a culture that supports ethical statistical practice."). Finally, the Appendix outlines responsibilities of employers (organizations and institutions): "Whenever organizations and institutions design the collection of, summarize, process, analyze, interpret, or present, data; or develop and/or deploy models or algorithms, they have responsibilities to use statistical practice in ways that are consistent with these Guidelines, as well as promote ethical statistical practice".

KSA 4: What are the alternative actions this ethical issue leads you to consider?

In general, as discussed earlier, there are always three actions to consider:

- Do nothing;
- Engage a colleague to formulate a response;
- Report the client/collaborator.

These are generic, so considering them with either the details of the case or in the context of the ASA GLs can help make them more specific – and so, more actionable.

A. **Professional integrity and accountability:** Engage a colleague to help you formulate a response – which may include reporting the client's behavior.

B. **Integrity of data and methods:** Engage a colleague for formulate a response – which may include reporting the client's behavior, particularly if harms may accrue to vulnerable populations due to inappropriate conclusions that can arise from inappropriate/not transparent reporting.

C. **Responsibilities to stakeholders:** The client who does not support valid inferences and transparency also does not support good science in general, and the ethical statistical practitioner cannot be prevented from fulfilling their obligation to science and the public (and the funder, if relevant) by the decisions or inclinations of the client. Reporting the client would alert the scientific community and potential funders to this client's willingness to engage in detrimental practices. Engaging with a colleague to discuss and formulate a response (such as reporting this behavior) would support Principle C.

D. **Responsibilities to research subjects, data subjects, or those directly affected by statistical practices:** Engage a colleague for formulate a response – which may include reporting the client's behavior, particularly if harms may accrue to vulnerable populations due to inappropriate conclusions that can arise from inappropriate/not transparent reporting.

E. **Responsibilities to members of multidisciplinary teams:** Engaging with other colleagues on the team may support the formulation of the response to this behavior. Others may be interested in collaborating on a full report (see F) that the client is *not* involved with.

F. **Responsibilities to fellow statistical practitioners and the profession:** Engaging with colleagues may lead to a separate report, in which the client/collaborator is not involved, and that provides full and transparent documentation of the work that was done and the inferences that are valid. That report would fulfill F4 and may also accomplish F5.

G. **Responsibilities of leaders, supervisors, and mentors in statistical practice:** Note that consideration of all of the GL Principles suggests that you should "engage a colleague to formulate a response". The leader, supervisor or mentor might be that colleague who is engaged! Practitioners who are not in leadership/mentorship or supervisory roles should consider that their leaders/supervisors/mentors are actually ethically obligated – just like the practitioner is – to follow the ASA Ethical Guidelines when practicing statistics. If the leader/supervisor/mentor is not themselves a practitioner, the Appendix outlines obligations of the non-statistical practitioner leader – in short, the colleague to engage could be a leader (but does not have to be). Also, the leader/supervisor/mentor might be the person to whom this detrimental collaborator is to be reported. It is worth noting that, having completed up to this point in the ethical reasoning process, you will be very well prepared to outline your concern, and why exactly you think the behavior is unethical (and/or detrimental). That will make it easier to report – as well as consult with colleagues or leaders about – the behaviors at issue.

H. **Responsibilities regarding potential misconduct:** The ethical statistical practitioner "Avoids condoning or appearing to condone statistical, scientific, or professional misconduct." (H2) Whether "detrimental research practices" are characterized as statistical, scientific, or professional misconduct, or all three types, the ethical statistical practitioner will ensure that their name is not associated with such practices. Withdrawing from the project is an alternative that, unless it is accompanied by some ancillary report of why they withdrew (see F) is only a slightly better option than "doing nothing". However, this may be the best alternative for the practitioner.

Summary of KSA 4: The client's behavior represents detrimental research practices that should not be endorsed and, if possible, should be prevented or offset of prevention is not possible. "Do nothing" is therefore not a plausible alternative; while "report the client" is an alternative that is consistent with most of the ASA GL Principles. However, the best supported alternative may be to collaborate with colleagues (excluding this client/collaborator) to ensure that a full and accurate report of the methods, data, and results becomes a part of the literature or documentation. This way, irrespective of whether the client's behavior is reported to (or that report is acted on by) funders or others in the community, the statistician has in fact discharged their responsibilities to subjects, science/the public and funders, other colleagues, and other statistical practitioners while also documenting the work in sufficient detail to offset the detrimental practices the client seeks to engage in. Note that the three alternatives we articulated for this case (Do nothing; Engage a colleague to formulate a response; Report the client/collaborator) were vague and very general, but we did manage to evaluate all three against the ASA GL Principles. From the list of how the three alternatives fit with the eight different ASA GL Principles, it can be seen that "do nothing" is contrary to all ASA Principles. Even though we started with very simplistic alternative responses, the analysis of whether/how the ASA GL Principles support any of them *was* informative.

KSA 5: Make and justify a decision

For this case, the decision that is best supported would be that the individual facing this kind of case/behavior from a client or collaborator should engage in conversations with a knowledgeable colleague to formulate an appropriate response. This response may include reporting the individual's behavior as detrimental research practices, but may also include formulating a full report, consistent with all ASA GL Principles about transparency, rigor and reproducibility, and the support of valid inferences, and ensuring that this "ancillary" report is also made available to anyone who can find/read the report the client seeks to complete/submit.

It can be difficult to "report" detrimental research practices – e.g., if you are not in an organization where "research" (as defined by the Federal Regulations) is taking place, or if detrimental research practices are not recognized as statistical and scientific misconduct. You may easily recognize behaviors as detrimental, but if leaders, supervisors and/or mentors do not recognize them as easily, it can be challenging to arrive at a satisfactory result (i.e., getting the detrimental practices stopped now, and prevented in future). Therefore, ensuring that a complete report is available to any interested reader might be the most feasible decision; collaborators on such a report may in fact strengthen it if they join the effort because they also perceive that the individual seeking to prevent full and transparent reporting is acting improperly. Justification for this choice of action includes the following:

Utilitarian perspective: Real harms accrue to science, the scientific community, and the public trust in science and the statistical/scientific professions when authors or presenters of "research" engage in inappropriate and detrimental research practices. These are recognized by the NASEM report and recommendations (2017), and are also included in the ASA GLs (H2) as inappropriate (contrary to the GLs). Statistical misconduct is highlighted in the NASEM report and has been the topic of many heated discussions about a "reproducibility crisis" in the US since about 2013, further emphasizing the real harms that accrue when the ASA GLs are not followed or when behaviors like this client's threaten to impede the ethical statistical practitioner in meeting their responsibilities. For these reasons, ensuring that a full and complete report becomes part of the scientific record must be considered at least as important as identifying detrimental practices – and those who seek to impede ethical statistical practice, whenever this is possible. Making this alternative report public (or as available as the inappropriate report is) may indirectly call out inappropriate behavior, but can help to directly offset harms that the inappropriate/incomplete report may have.

Virtue perspective: Following the ASA GLs is the responsibility of all ethical statisticians. In following all relevant principles of the ASA GLs, the ethical statistical practitioner "Avoids condoning or appearing to condone statistical, scientific, or professional misconduct. Encourages other practitioners to avoid misconduct or the appearance of misconduct." (H2). Further, those supervising, leading, and mentoring statistical practitioners also have the responsibility to "Ensure appropriate statistical practice that is consistent with these Guidelines. Protect the statistical practitioners who comply with these

Guidelines, and advocate for a culture that supports ethical statistical practice." (G1). The ethical statistical practitioner cannot allow the client's detrimental practice to impede *their* ethical practice, nor can they allow detrimental practices to cause them to appear to condone statistical, scientific, or professional misconduct. Thus, ensuring that a full and correct/transparent report becomes available to any reader of the incorrect report is an important response, but calling out this behavior as unacceptable – even if it is common – may be equally important under the virtue perspective.

KSA 6. Reflect on the decision

Reflecting on this decision can begin from any of the difficulties that arose during the entire ER process. For example, the observation that "do nothing" is both *a decision* and one that is inconsistent with ethical practice; but might also be the most appealing, least challenging option. It might be worth the time to consider the justification for "doing nothing" – it is easiest on the decider (you). But this decision does not address any of the practitioner's obligations and responsibilities, nor does it promote the practice of statistics or data science. In fact, the "benefit" (easy for me!) is significantly offset by the harms-to science, the public trust, and the practice of statistics. If you consider that **refusing to do nothing** in the face of detrimental or unethical behavior is a burden that, perhaps unfairly, seems to fall only on your shoulders, it must also be noted that the next person facing the same detrimental behavior will have the same decisions to make that you have had. The person who did nothing before this detrimental behavior became *your* problem made a decision that created the opportunity for you to have to face the choice, too. This is only one way that "doing nothing" harms the profession, by failing to offset detrimental practices with ethical responses. If it seems that engaging with a colleague or reporting the detrimental practitioner are beyond your capabilities, maybe you can think of another action – revisit KSA 4 and 5 – that is more than "nothing", but still lowers the likelihood of this detrimental practice in the future.

Case 2B, ACM CE. *Client/employer has not identified important confounders or sources of bias in the data they are collecting. Their data collection mechanisms are already deployed, and you are committed to creating a system to execute the analysis of the data.*

KSA 3: What is the ethical issue this client's behavior raises (according to the ACM CE)?

ACM CE 1: The ethical computing has an obligation to avoid harm, and "should follow generally accepted best practices unless there is a compelling ethical reason to do otherwise. Additionally, the consequences of data aggregation and emergent properties of systems should be carefully analyzed." Also, "computing professionals should *take action to avoid creating systems or technologies that disenfranchise or oppress people*. Failure to design for inclusiveness and accessibility may constitute unfair discrimination."

ACM CE 2: "Computing professionals should respect the right of those involved to transparent communication about the project. Professionals should be cognizant of any serious negative consequences affecting any stakeholder that may result from poor quality work and should resist inducements to neglect this responsibility."

<nothing particular from CE 3>

ACM CE 4: failures to comply with the CE are the responsibility of all ethical computing professionals, because they have an obligation to "Uphold, promote, and respect the principles of the Code." (ACM 4.1).

Summary of KSAs 1-3: the client/employer in this vignette is themselves violating several principles and guidelines from the ACM CE, and moreover seeks to cause the computing professional to violate the CE as well. Not only does this behavior violate the CE it also may compound that by causing others to violate them as well, propagating unethical behavior and potentially creating a "new norm" for this particular violation.

KSA 4: What are the alternative actions this ethical issue leads you to consider?

ACM CE 1: "If leaders do not act to curtail or mitigate such risks, it may be necessary to "blow the whistle" to reduce potential harm."

ACM CE 2: "In cases where misuse or harm are predictable or unavoidable, the best option may be to not implement the system."

ACM CE 3: "Leaders should pursue clearly defined organizational policies that are consistent with the Code and effectively communicate them to relevant stakeholders."

ACM CE 4: The behavior should be actively resisted (i.e., the opposite of "do nothing"), and the client informed of their actions' violation of the CE. Failing to discourage the client from continuing to violate –and inducing others to violate –the ACM CE compounds the client's violation and adds to it the computing professional's! So, simple withdrawal/refusal is insufficient.

Summary of KSA 4: The best supported alternative comes from ACM CE 3: "Leaders should pursue clearly defined organizational policies that are consistent with the Code and effectively communicate them to relevant stakeholders." That means the individual facing this kind of case/behavior from a client (or employer) should engage in conversations with someone in a leadership position (probably at a higher level <but peers, if empowered to take action, can help as well> if their direct employer is the one trying to cause this CE violation). The *leaders'* responsibility is to communicate how the policy – in this case, to follow the ACM CE and not agree to this client's behavior – prevents compliance with the request. The leaders may never find out about such behavior if the employees do not engage in open discussions (as the ACM CE suggests all should do!) Note that the three alternatives we articulated for this case (Do nothing/allow client's perspective and do not report;

Engage a colleague to formulate a response; Refuse to continue on the project) were vague and very general, but we did manage to evaluate all three against the ACM CE Principles. We found that "do nothing" was contrary to all ACM Principles, and also that even though we started with very simplistic alternative responses, the analysis of whether/how the ACM CE Principles support any of them *was* informative.

Several ACM CE principles note that it may be necessary to blow the whistle, making that a clear option for how to respond. They also state that the best option (i.e., alternative action) may be to not implement the system. Obviously, neither of these alternatives resembles "do nothing"! Also, it is possible that for YOU to "not implement the system" is effectively you simply withdrawing (which was one of the alternatives we set up for evaluation in the last chapter); we saw that all of the CE principles only partially support this alternative, because all of the CE principles include some kind of statement about notifying the client (that they are in violation of the ACM CE) and also notifying others of the harms that the system may create. So "refuse to continue" or "do not implement the system" are *not* actually supported alternative actions. Instead, a third alternative (based on the tables in the previous 2 chapters) is to "document fully why the system should not be implemented/why you are withdrawing, and make sure that the client and your colleagues (and possibly the public) are aware of the justification as well as the fact that you are withdrawing/refuse to implement the system". This may seem a lot like "whistleblowing", but that term is usually reserved for reporting illegal, rather than unethical, behaviors. For example, the US Federal office, Occupational Safety and Health Administration (OSHA) maintains extensive information about whistleblowing: "OSHA's Whistleblower Protection Program enforces the whistleblower provisions of more than twenty whistleblower statutes protecting employees who report violations of various workplace safety and health, airline, commercial motor carrier, consumer product, environmental, financial reform, food safety, health insurance reform, motor vehicle safety, nuclear, pipeline, public transportation agency, railroad, maritime, and securities laws." (For more information about whistleblowing, and particularly, laws that protect the whistleblower, please seek expert and authoritative guidance!)

KSA 5: Make and justify a decision

For Case 2B the decision that is best supported would be that the individual facing this kind of case/behavior from a client (or employer) should engage in conversations with someone in a leadership position to formulate a response that is consistent with the ACM CE. Justification for this choice of action includes the following:

Utilitarian perspective: harms may accrue to unknown individuals if a system is developed and deployed with biases in the data that are not addressed, and ultimately to the profession. There are no benefits to any stakeholders if the data are used without adjustments for bias – which can take the form of simulations and ongoing monitoring (or, could involve collecting different or additional data). The benefit to the client/employer is that they do not need to change how their data collection happens, with potential harms accruing to them like additional costs to collect more/different data. However, other potential harms also accrue given a potentially biased dataset, and these may offset any benefits.

Virtue perspective: Following the ACM CE requires both that the computing professional complies and that the computing professional encourages and promotes the client/employer to also comply. The most efficient method to accomplish both would be to communicate openly with an organizational leader to communicate a policy of following the ACM CE to the client/employer. All of the ACM CE principles support taking action and not: a) doing nothing; or b) simply withdrawing, potentially allowing another computing professional to execute the request without ensuring that the system is developed according to the ACM CE.

KSA 6: Reflecting on the decision

KSA 6, reflecting on the decision, may seem superfluous because you've made and justified your decision, and you should just move on at that point. However, thinking about what was hard about making the decision, what additional information would have been helpful, how to get better at these challenging features of ethical reasoning, how to help create the culture that promotes fluency in ethical reasoning are all examples of reflections. Again, these may seem like "an extra step" for those new to this kind of reasoning, but this kind of reflection will in fact support a more ethical workplace in the future. Another type of reflection may seem much less superfluous because it invites you to consider your decision (and its justification) from the perspective of others: Stadler (1986) proposed three "tests" for your decision that are derived from the writings of ethicist/philosopher Immanuel Kant. The tests are for the universality, justice, and publication of the end result of your ethical reasoning, your decision. The test of *universality* is, "would this decision work in every other similar situation?" The test of *justice* is, "does this decision affect all people equally? Would this be a fair "law" if it was enacted?" and the test of publication (originally "publicity") is, "would I like it if everyone found out that I made this decision?" (Would you want your decision published or made public in/on the news?)

Thinking about the next person who has to think through this same decision or a similar one, what additional information would be/would have been helpful to you (was there additional prerequisite knowledge you wish you had)? Thinking about your experience reasoning your way to an ethical decision, does a decision like the one you thought through help to create the culture that promotes fluency in ethical reasoning and/or a more ethical workplace?

Obviously, the steward wants the next person who has to make this or a similar decision to think it through and choose the ethical, stewardly option. Reflecting on your decision and decision-making process may increase the likelihood of the next decision maker doing so in a stewardly way.[21] Acting to eliminate detrimental research practices from any practitioners' consideration is not an explicit responsibility according to the ACM CE, but it is *implicit*. Thus, one might encourage the creation of (or create) materials for their organization or team introducing the ACM CE, but ensuring that detrimental practices are specifically called out as unethical. One of the purposes of this book is to ensure that everyone recognizes the importance and utility of open discussions of what to do/how to ensure that the ASA GLs, or ACM CE (or both, when applicable) are followed. Such materials would certainly strengthen the professionalism of new practitioners, and might be useful to new/potential clients. The decision to promote –and engage in – open discussions about ethical issues and compliance with the ACM CE passes the "publication" test. Acting to promote engagement in such conversations throughout the organization in the future passes the "universality" test. Supporting a transparent and accountable ethical decision-making process for all computing and quantitative professionals satisfies the "justice" test.

Notes on KSA 6, reflection

One key purpose of "reflection" as a final step in the process of ethical reasoning is to evaluate the analysis you have (or someone else has) done, to determine if the analysis yields the decision that is consistent with either the ASA or ACM perspectives – or with both (or neither). If your workplace has ethical guidelines or codes of conduct, you might carry out the ethical reasoning process and then reflect on how the ASA or ACM perspectives are, or are not, represented in your workplace ethical practice standards. Promoting ethical quantitative practice can involve making changes to workplace ethical standards as well as implementing the professional practice standards in your own practice.

Both the ASA and ACM note that their practice standards are "guides" and not rules; neither organization believes that there is one simple answer to any given question. The ACM CE specifically notes that "When thinking through a particular issue, a computing professional may find that multiple principles should be taken into account, and that different principles will have different relevance to the issue. Questions related to these kinds of issues can best be answered by thoughtful consideration of the fundamental ethical principles, understanding that the public good is the paramount consideration."

---

[21] This paragraph excerpted from Tractenberg, RE. (2019-b, May 1). Strengthening the practice and profession of statistics and data science using ethical guidelines. Retrieved from https://doi.org/10.31235/osf.io/93wuk

(Preamble), while in Principle 2 the ACM CE states that "Professional competence also requires skill in communication, in reflective analysis, and in recognizing and navigating ethical challenges." (2.2). Thus, the ethical reasoning KSA for reflection (KSA 6) is echoed in the ACM CE. KSA 6 is described as the final "step" in ethical reasoning so that you can "reflect" on both the decision/its justification and the given analysis overall – to determine how best to notify others of the decision and its justification or how to act so that such choices are not faced by others in the future. Note that, while the ER process ends with reflection, the ACM CE features reflection throughout the general analysis of which principles to follow and how to apply the CE. The ASA GLs do not discuss reflection specifically. It is important to note that the process of reflection in the ER steps is intentionally at the end, but the careful consideration that the CE and GLs both state the ethical practitioner will do as they make ethical decisions – which the CE calls "reflection", but also refers to as "analysis" – is embedded within the other ER steps. That is, the CE and ER steps do not disagree about the role of reflection; rather, the differences in how they use the term comes from the specific use of the term in the ER process and the more general use of the term in the CE.

Discussion Questions:

Complete the Ethical Reasoning process for the two cases from the preceding chapters, utilizing the ASA GLs (Cases 1A, 2A) and ACM (Cases 1B, 2B).

1. Case 1A was presented using the ASA GLs; do that analysis using the ASA GLs on Case 2A, "Client/collaborator is not aware of statistical concepts of sampling, bias, etc. and may not be able to identify important confounders or sources of bias. Their data are already collected/data collection mechanisms are already deployed, and you are committed to planning and executing the analysis." Discuss similarities and differences between results on the ASA-driven cases.

2. Case 2B was presented using the ACM CE; do that analysis using the ACM CE on Case 1B, "The client/collaborator does not support/is not committed to the correct, transparent reporting and documentation of the system, model, or algorithm under development." Discuss similarities and differences between results on the ACM-driven cases.

Section 2. Establishing familiarity with ASA and ACM principles/elements as they relate to the seven tasks of the statistics and data science pipeline: anticipating what problems may arise

Chapter 2.1
Introduction to Section 2

The emphasis in Section 2 is on your learning how to do the stakeholder analysis (SHA), and how the SHA plus the CE/GLs can provide guidance for your thinking/planning/workflows, as well as for each of the typical tasks in statistics and data science. The seven typical activities of the quantitative practitioner represent the *statistics and data science pipeline*. As promised, ethical reasoning is a lot more than simply remembering our prerequisite knowledge (KSA 1), and even more complicated than organizing it according to our decision-making perspectives (virtue/utilitarian; KSA 2). Although remembering and understanding these ethical practice standards are situated at the "lowest complexity levels" in Bloom's taxonomy, with so much material – and much of it fairly abstract – developing sufficient prerequisite knowledge is *not simple*! An excellent way to ensure that Bloom's 1-2 (remembering, understanding) capabilities are firmly established is to practice *applying* that knowledge (Bloom's 3) and even using it to make predictions and figure out what you should do when the knowledge must be brought to bear (Bloom's 4-5). So, Section 2 is all about reinforcing the prerequisite knowledge, but also developing your confidence and capabilities in applying your Bloom's levels 3-5 capabilities to that knowledge.

Section 2 is designed to reinforce your prerequisite knowledge (KSA 1) and deepen your understanding of how decision-making frameworks function in day-to-day working life (KSA 2); these are how you will use the ethical practice standards 99% of the time, so ensuring you are confident in your abilities with KSAs 1-2 is one objective of Section 2. Although KSAs 1 and 2 are essential whenever you are practicing, the other KSAs are more important when you also have something going on in KSA 3: namely, when you can/do identify an ethical challenge. The chapters in Section 2 are separated into the seven typical activities of the quantitative practitioner, also called the *statistics and data science pipeline*:

Planning/Designing
Data collection/munging/wrangling
Analysis (perform or program to perform)
Interpretation
Documenting your work
Reporting your results/communication

Engaging in team science/teamwork

The learning objectives associated with our discussion of these seven activities
are to:

- describe how different individuals ("stakeholders") may be affected by
  decisions and actions;
- enumerate harms and benefits that are most clearly relevant for each
  stakeholder with respect to the activity; and
- identify which ASA GL and ACM CE Principles (and/or specific
  elements) seem most relevant to each activity/task.
- Consider whether ASA and ACM have different perspectives on how
  to accomplish each task in the most ethical manner.

Part of the "prerequisite knowledge" is recognizing which ones are most
relevant, but the fact that multiple aspects of each (GL/CE) pertain in any given
situation underscores the fact that **guidance is available** from these sources.

The following chapters offer brief reviews of those ASA GL and ACM CE
principles - and within principles, which elements- that are specific or
indirect/oblique in their instructions about how the ethical practitioner
accomplishes typical tasks in quantitative practice.

The eight Principles of the ASA GLs are emphasized in these brief analyses,
excluding the Appendix because it is specific to organizations and institutions.
While the seven tasks in the pipeline are carried out in the organizational or
institutional context, the Principles are most directly relevant to the statistical
practitioner (which is why they're emphasized). Similarly, the two elements of
the ACM CE that are focused on membership (Principle 4, elements 4.1 and 4.2)
are more oriented towards the ACM's organizational priorities, rather than
ethical practice, so ACM Principle 4 is also excluded from these analyses.

Both the ASA and ACM ethical practice standards feature guidance for leaders
or those practitioners who are in leadership roles (ASA Principle G, and ACM
Section 3). These are included here in our considerations of the pipeline tasks
principally to help the reader identify attributes of the workplace that will be
most supportive of ethical practice. However, familiarity with the relationships
between leadership responsibilities to promote and engage in ethical practice
in statistics, data science, and computing can also help readers develop into the
most ethical type of leader possible.

Chapter 2.2
Planning/Designing

ASA Ethical Guidelines (GLs) for Statistical Practice: on planning/designing

This is a brief review of those ASA GL principles - and within principles, which elements- that are specific or indirect/oblique in their instructions about how the ethical practitioner plans or designs projects or other work products.

Overall: The first paragraph of the preamble to the GLs states that:

> "The American Statistical Association's Ethical Guidelines for Statistical Practice are intended to help statistical practitioners make decisions ethically. In these Guidelines, "statistical practice" includes activities such as: *designing the collection of,* summarizing, processing, analyzing, interpreting, or presenting, data; as well as *model or algorithm development* and deployment. Throughout these Guidelines, the term "statistical practitioner" includes all those who engage in statistical practice, regardless of job title, profession, level, or field of degree. The Guidelines are intended for individuals, but these principles are also relevant to organizations that engage in statistical practice." (Emphasis added)

These comments appear in the "Purpose of the Guidelines" and are intended to communicate that the ethical practitioner strives to plan/design - and supports the design – of work that is based on, and promoting, transparent, reproducible, and valid decisions. Moreover, the ethical practitioner both acts in a manner consistent with the GLs in their own project/program design or planning, and also encourages others to do so. Planning of every project must include considerations of the potential harm that a study, analysis, or other work product can bring or create. In experimental or clinical trial design, this is a consideration for power and sample size calculation, while in other types of analyses, unethical sourcing of data must be considered.

Principle A: Professional integrity and accountability

Professional integrity and accountability require taking responsibility for one's work. Ethical statistical practice supports valid and prudent decision making with appropriate methodology. The ethical statistical practitioner represents their capabilities and activities honestly, and treats others with respect.

The ethical statistical practitioner:

A1 Takes responsibility for evaluating potential tasks, assessing whether they have (or can attain) sufficient competence to execute each task, and that the work and timeline are feasible. Does not solicit or deliver work for which they are not qualified, or that they would not be willing to have peer reviewed.

A2 Uses methodology and data that are valid, relevant, and appropriate, without favoritism or prejudice, and in a manner intended to produce valid, interpretable, and reproducible results.

A3 Does not knowingly conduct statistical practices that exploit vulnerable populations or create or perpetuate unfair outcomes.

A4 Opposes efforts to predetermine or influence the results of statistical practices, and resists pressure to selectively interpret data.

A6 Strives to follow, and encourages all collaborators to follow, an established protocol for authorship. Advocates for recognition commensurate with each person's contribution to the work. Recognizes that inclusion as an author does imply, while acknowledgement may imply, endorsement of the work.

A7 Discloses conflicts of interest, financial and otherwise, and manages or resolves them according to established policies, regulations, and laws.

A9 Takes appropriate action when aware of deviations from these Guidelines by others.

Principle A, as well as these seven identified elements, relate specifically to how important it is for the ethical practitioner to plan for/design competently and transparently, using appropriate methodology and not limited or restricted by favoritism or prejudice. Even if workplace structures mean that the individual who plans or designs a study/project/program will not be the one executing the design, the ethical practitioner plans/designs ethical work.

Principle B. Integrity of data and methods

The ethical statistical practitioner seeks to understand and mitigate known or suspected limitations, defects, or biases in the data or methods and communicates potential impacts on the interpretation, conclusions, recommendations, decisions, or other results of statistical practices.

The ethical statistical practitioner:

B1 Communicates data sources and fitness for use, including data generation and collection processes and known biases. Discloses and manages any conflicts of interest relating to the data sources. Communicates data processing and transformation procedures, including missing data handling.

B2   Is transparent about assumptions made in the execution and interpretation of statistical practices including methods used, limitations, possible sources of error, and algorithmic biases. Conveys results or applications of statistical practices in ways that are honest and meaningful.

B3   Communicates the stated purpose and the intended use of statistical practices. Is transparent regarding a priori versus post hoc objectives and planned versus unplanned statistical practices. Discloses when multiple comparisons are conducted, and any relevant adjustments.

B4   Meets obligations to share the data used in the statistical practices, for example, for peer review and replication, as allowable. Respects expectations of data contributors when using or sharing data. Exercises due caution to protect proprietary and confidential data, including all data that might inappropriately harm data subjects.

B5   Strives to promptly correct substantive errors discovered after publication or implementation. As appropriate, disseminates the correction publicly and/or to others relying on the results.

B6   For models and algorithms designed to inform or implement decisions repeatedly, develops and/or implements plans to validate assumptions and assess performance over time, as needed. Considers criteria and mitigation plans for model or algorithm failure and retirement.

B7   Explores and describes the effect of variation in human characteristics and groups on statistical practice when feasible and relevant.

Every element in Principle B, as well as the Principle itself, relates directly or indirectly to planning and design practices. Direct relevance to the plan includes knowing ahead of time whether there are known or suspected defects/limitations or biases in the data that can have an effect in later use of that data; this is clearly important for the design/planning of ethical projects and analyses. More indirectly, because ending up with reproducible "interpretation, conclusions, recommendations, decisions, or other results of statistical practices" requires that planning is careful, professional, and ethical; thus, plans and designs relating to the data as well as the methods must all keep these endpoints in mind. This includes having a plan for what to do in case any error is (eventually) discovered. The ethical practitioner bears responsibility for the integrity of data and methods in the planning stages and beyond.

Principle C. Responsibilities to stakeholders

The ethical statistical practitioner:

C1   Seeks to establish what stakeholders hope to obtain from any specific project. Strives to obtain sufficient subject-matter knowledge to conduct meaningful and relevant statistical practice.

C3   Uses practices appropriate to exploratory and confirmatory phases of a project, differentiating findings from each so the stakeholders can understand and apply the results.

C4   Informs stakeholders of the potential limitations on use and re-use of statistical practices in different contexts and offers guidance and alternatives, where appropriate, about scope, cost, and precision considerations that affect the utility of the statistical practice.

C5   Explains any expected adverse consequences from failing to follow through on an agreed-upon sampling or analytic plan.

C7   Understands and conforms to confidentiality requirements for data collection, release, and dissemination and any restrictions on its use established by the data provider (to the extent legally required). Protects the use and disclosure of data accordingly. Safeguards privileged information of the employer, client, or funder.

In both the planning and execution of those plans, the ethical practitioner is aware of, and prepared for, contingencies that may affect the wider scientific community and the public, as well as the funder and/or client. The ethical practitioner has responsibilities to plan carefully for the final disposition of the results of the project, as well as how the data may need to be maintained (e.g., securely for a pre-specified period of time, and then destroyed). These five elements of Principle C are specifically supportive of ethical planning and design.

Principle D. Responsibilities to research subjects, data subjects, or those directly affected by statistical practices

The ethical statistical practitioner does not misuse or condone the misuse of data. They protect and respect the rights and interests of human and animal subjects. These responsibilities extend to those who will be directly affected by statistical practices.

The ethical statistical practitioner:

D1   Keeps informed about and adheres to applicable rules, approvals, and guidelines for the protection and welfare of human and animal subjects. Knows when work requires ethical review and oversight.

D2   Makes informed recommendations for sample size and statistical practice methodology in order to avoid the use of excessive or inadequate numbers of subjects and excessive risk to subjects

D3   For animal studies, seeks to leverage statistical practice to reduce the number of animals used, refine experiments to increase the humane treatment of animals, and replace animal use where possible.

D4   Protects people's privacy and the confidentiality of data concerning them, whether obtained from the individuals directly, other persons, or existing records. Knows and adheres to applicable rules, consents, and guidelines to protect private information.

D5   Uses data only as permitted by data subjects' consent when applicable or considering their interests and welfare when consent is not required. This includes primary and secondary uses, use of repurposed data, sharing data, and linking data with additional data sets.

D6   Considers the impact of statistical practice on society, groups, and individuals. Recognizes that statistical practice could adversely affect groups or the public perception of groups, including marginalized groups. Considers approaches to minimize negative impacts in applications or in framing results in reporting.

D7   Refrains from collecting or using more data than is necessary. Uses confidential information only when permitted and only to the extent necessary. Seeks to minimize the risk of re-identification when sharing de-identified data or results where there is an expectation of confidentiality. Explains any impact of de-identification on accuracy of results.

D8   To maximize contributions of data subjects, considers how best to use available data sources for exploration, training, testing, validation, or replication as needed for the application. The ethical statistical practitioner appropriately discloses how the data is used for these purposes and any limitations.

D9   Knows the legal limitations on privacy and confidentiality assurances and does not over-promise or assume legal privacy and confidentiality protections where they may not apply.

D10   Understands the provenance of the data, including origins, revisions, and any restrictions on usage, and fitness for use prior to conducting statistical practices.

D11   Does not conduct statistical practice that could reasonably be interpreted by subjects as sanctioning a violation of their rights. Seeks to use statistical practices to promote the just and impartial treatment of all individuals.

Ethical obligations to research subjects are numerous in the planning stages of any project or study, which is why every one of Principle D elements, and the overall Principle itself, are highlighted here. As noted earlier, these are relevant for sample size calculations and other design considerations, and elements D2, D3, D5 and D7 in particular address whether appropriate plans had been in place for a data set that is under consideration for collection or use. If appropriate approvals had not been obtained, then the data are considered unethically sourced, and ethical practice suggests not using that data (i.e., planning to use other data). More specifically, planning a study/report/project that might create the impression of violating the rights of those whose data are to be used is unethical. Not only does the ethical practitioner make alternate plans (for different, ethically sourced, data), but s/he also acts to discourage others from using such unethically sourced data.

Principle E. Responsibilities to members of multidisciplinary teams

The ethical statistical practitioner:

E4 Avoids compromising validity for expediency. Regardless of pressure on or within the team, does not use inappropriate statistical practices.

Whether or not the team is engaged in research or non-science work, the ethical practitioner ensures that what will be reported was ethically designed and executed. The ethical obligations of the statistical practitioner are embedded throughout practice, partly to discharge our Responsibilities to research subjects, data subjects, or those directly affected by statistical practices (D) to our team members (E), and to the profession (F). While some non-statistical team members may seek to ensure that a publication or other countable work product results from the team's effort –without consideration for ethical practice parameters for their or the statistician's field - responsibilities to the research subjects (D) and other statisticians/statistical practitioners (F, next Principle), preclude research team colleagues from promoting unethical planning or design. It is precisely these responsibilities to others that require the ethical practitioner to ensure that all plans and designs are consistent with E4.

Principle F. Responsibilities to fellow statistical practitioners and the Profession

The ethical statistical practitioner:

F2 Helps strengthen, and does not undermine, the work of others through appropriate peer review or consultation. Provides feedback or advice that is impartial, constructive, and objective.

F5 Serves as an ambassador for statistical practice by promoting thoughtful choices about data acquisition, analytic procedures, and data structures among non-practitioners and students. Instills appreciation for the concepts and methods of statistical practice.

Ethical planning, and planning ethical projects/studies, are embedded in our responsibilities to other practitioners and the profession itself. Ensuring results that are useable and reproducible is an essential motivator for ethical planning. This may come in the form of reviewing plans by others, where F2 states the feedback and advice should be impartial, constructive and objective.

Principle G. Responsibilities of leaders, supervisors, and mentors in statistical practice

Those leading, supervising, or mentoring statistical practitioners are expected to:

G1 Ensure appropriate statistical practice that is consistent with these Guidelines. Protect the statistical practitioners who comply with these Guidelines, and advocate for a culture that supports ethical statistical practice.

G5 Establish a culture that values validation of assumptions, and assessment of model/algorithm performance over time and across relevant subgroups, as needed. Communicate with relevant stakeholders regarding model or algorithm maintenance, failure, or actual or proposed modifications.

The ethical statistical practitioner follows the ASA GLs relating to planning and design, whether as a practitioner or as a leader, supervisor, or mentor. When ethical practice is not supported in the workplace, the ethical practitioner recognizes that this lack of support for ethical practice is contrary to the GLs (Principle G and the Appendix), and should not be allowed/followed. By encouraging the organization/institution to uphold, or permit them to uphold, the ASA Ethical GLs for Statistical Practice, the ethical practitioner directly supports the freedom and responsibility of statistical practitioners who comply with these GLs.

Principle H. Responsibilities regarding potential misconduct

The ethical statistical practitioner:

> H2 Avoids condoning or appearing to condone statistical, scientific, or professional misconduct. Encourages other practitioners to avoid misconduct or the appearance of misconduct.

The ethical practitioner engages in, and directly and indirectly supports, plans and designs of ethical studies, projects, and programs so as to avoid condoning incompetent and unprofessional behavior or misconduct. While not illegal, prioritizing expedience over validity (contrary to E3) is an example of statistical and scientific misconduct.

ACM Code of Ethics: on planning/designing

Overall: the first sentence of the Preamble states,
Computing professionals' actions change the world. To act responsibly, they should reflect upon the wider impacts of their work, consistently supporting the public good.
And in the second paragraph it states that the Code is
...based on the understanding that the public good is always the primary consideration.

In their planning and designing, the ethical computing professional should always keep the public good, and the minimization of harms, as their primary focus.

1. General Ethical Principles

> 1.1 An essential aim of computing professionals is to minimize negative consequences of computing, including threats to health, safety, personal security, and privacy. ... and Computing professionals should consider whether the results of their efforts will respect diversity, will be used in socially responsible ways, will meet social needs, and will be broadly accessible.

1.2   Avoiding harm begins with careful consideration of potential impacts on all those affected by decisions. When harm is an intentional part of the system, those responsible are obligated to ensure that the harm is ethically justified. In either case, ensure that all harm is minimized.

To minimize the possibility of indirectly or unintentionally harming others, computing professionals should follow generally accepted best practices unless there is a compelling ethical reason to do otherwise. Additionally, the consequences of data aggregation and emergent properties of systems should be carefully analyzed.

A computing professional has an additional obligation to report any signs of system risks that might result in harm. If leaders do not act to curtail or mitigate such risks, it may be necessary to "blow the whistle" to reduce potential harm. However, capricious or misguided reporting of risks can itself be harmful. Before reporting risks, a computing professional should carefully assess relevant aspects of the situation.

1.4   Technologies and practices should be as inclusive and accessible as possible and computing professionals should take action to avoid creating systems or technologies that disenfranchise or oppress people. Failure to design for inclusiveness and accessibility may constitute unfair discrimination.

1.6   Computing professionals should establish transparent policies and procedures that allow individuals to understand what data is being collected and how it is being used, to give informed consent for automatic data collection, and to review, obtain, correct inaccuracies in, and delete their personal data.

In their planning and designing, the ethical computing professional considers the potential for risks of harms as well as harms, and ensures that systems are designed with minimization of harms/risks, and methods for remedying bias and/or harms that may arise in the future (as a result of a system they planned/designed/developed).

2. Professional Responsibilities

2.4   Whenever appropriate, computing professionals should seek and utilize peer and stakeholder review.

2.5   Computing professionals are in a position of trust, and therefore have a special responsibility to provide objective, credible evaluations and testimony to employers, employees, clients, users, and the public. Computing professionals should strive to be perceptive, thorough, and objective when evaluating, recommending, and presenting system descriptions and alternatives. Extraordinary care should be taken to identify and mitigate potential risks in machine learning systems. A system for which future risks cannot be reliably predicted requires frequent reassessment of risk as the system evolves in use, or it should not be deployed. Any issues that might result in major risk must be reported to appropriate parties.

2.6 A computing professional is responsible for evaluating potential work assignments. This includes evaluating the work's feasibility and advisability, and making a judgment about whether the work assignment is within the professional's areas of competence.

2.8 computing professionals should not access another's computer system, software, or data without a reasonable belief that such an action would be authorized or a compelling belief that it is consistent with the public good.

2.9 Robust security should be a primary consideration when designing and implementing systems. Computing professionals should perform due diligence to ensure the system functions as intended, and take appropriate action to secure resources against accidental and intentional misuse, modification, and denial of service.

The ethical computing professional seeks expert peer review of/input into their planning and designing; they fully explore what their system would do/be able to do, and what other systems or data their system would use or access. Systems must be designed to be robust, while also allowing remedies if misuses or other harmful applications of their work arise in future.

3. Professional Leadership Principles

3.1 People—including users, customers, colleagues, and others affected directly or indirectly— should always be the central concern in computing. The public good should always be an explicit consideration when evaluating tasks associated with research, requirements analysis, design, implementation, testing, validation, deployment, maintenance, retirement, and disposal. Computing professionals should keep this focus no matter which methodologies or techniques they use in their practice.

3.4 Designing or implementing processes that deliberately or negligently violate, or tend to enable the violation of, the Code's principles is ethically unacceptable.

3.5 Computing professionals should be fully aware of the dangers of oversimplified approaches, the improbability of anticipating every possible operating condition, the inevitability of software errors, the interactions of systems and their contexts, and other issues related to the complexity of their profession—and thus be confident in taking on responsibilities for the work that they do.

3.6 Leaders should take care when changing or discontinuing support for system features on which people still depend. Leaders should thoroughly investigate viable alternatives to removing support for a legacy system.

The ethical computing professional, particularly when in leadership positions, ensure that planning and designing of systems keeps the public good as a focal objective. This includes considerations of whether a system might be/become part of the infrastructure of other organizations and also whether or not any aspect of the system being planned/designed might violate, or enable the violation of, any of the ACM Code principles. The stakeholder analysis (SHA) table below shows some harms, benefits, unknown outcomes, and unknowable outcomes, relating to following the GLs and CE in planning and designing.

Potential result: Stakeholder:	HARM	BENEFIT	UNKNOWN	UNKNOWABLE
YOU	Takes time, may incur liability	Ensures compliance with CE/GL		
Your boss/client	Takes time	Ensures compliance		
Unknown individuals	If ineffective, can create bias or privacy breaches	Mitigates risk of bias/privacy breach/harms		Not all risks/harms can be foreseen and mitigated
Employer	Takes time	Supports compliance		Could limit business/profit
Colleagues	May complicate colleagues' work (challenging them to follow GL/CE)	Supports compliance		Commitment to GL/CE strengthens colleagues' trust in ethical practice
Profession	If not competently done, undermines the profession	Supports trust in the profession		Commitment to GL/CE strengthens trust in profession and enables solutions/ conversations
Public/public trust	If ineffective, can undermine public trust in systems and the profession	Supports public trust in the profession	Transparency supports trust, even if misuse does occur	Engagement with GL/CE promotes trust even if misuse does occur

Table 2.2.1. *Stakeholder Analysis template: Planning/designing*

Questions for Discussion:

Discuss similarities and differences between the ASA GLs and ACM CE principles that are discussed above, in terms of how they apply to planning and designing activities.

What are the most striking similarities (e.g., what is unexpectedly similar)? Are there particular differences that stand out because they are A) highly predictable? B) unexpected? C) have different benefits or harms to different stakeholders?

Rank GL/CE Principles from most to least important to planning and design, using their own language or your own perspective (and identify which you chose in your answer). What does the rankability mean for you or for the discipline of statistics and data science?

Can following the GL/CE affect how you plan or design projects in your typical work? How/How not?

Is the SHA in Table 2.1.1 enough to ensure ethical practice? Why/why not?

Discuss how the ASA and ACM are, or are not, fully aligned with current "policies regarding human subjects, live vertebrate animal subjects in research, and safe laboratory practices".

Discuss how the ASA and ACM are, or are not, fully aligned with current policies (for your institution, organization, or lab/work team) on "data acquisition and laboratory tools; management, sharing and ownership".

Chapter 2.3
Data collection/munging/wrangling

ASA Ethical Guidelines (GLs) for Statistical Practice: on data collection/munging/ wrangling

Overall: The first paragraph of the preamble to the GLs states that:

"In these Guidelines, "statistical practice" includes activities such as: *designing the collection of, summarizing, processing,* analyzing, interpreting, or presenting, data; as well as model or algorithm development and deployment." And the final sentence says, "To justify unethical behaviors, or to exploit gaps in the Guidelines, is unprofessional, and inconsistent with these Guidelines." (Emphasis added)

These comments in the "Purpose of the Guidelines" are intended to communicate that the ethical practitioner strives to do, and supports the design of, work that is based on data obtained in good faith, and not obtained unethically. Moreover, the ethical practitioner both acts in a manner consistent with the GLs in their collection and munging/wrangling of data, and does not exploit gaps in the GLs in cases where there may be less specific guidance about data collection, munging, or wrangling. In experimental studies or clinical trials, data must be collected with informed consent (from human participants) and in an ethical and humane way (from human and animal participants), while in other types of analyses or projects, unethically sourced data must be avoided. Unethically collecting data is professional misconduct.

Principle A: Professional Integrity and Accountability

The ethical statistical practitioner:

A2 Uses methodology and data that are valid, relevant, and appropriate, without favoritism or prejudice, and in a manner intended to produce valid, interpretable, and reproducible results.

A3 Does not knowingly conduct statistical practices that exploit vulnerable populations or create or perpetuate unfair outcomes.

A4 Opposes efforts to predetermine or influence the results of statistical practices, and resists pressure to selectively interpret data.

These three identified elements of Principle A relate to the importance of ethical data collection, munging, and wrangling. The ethical practitioner collects valid, relevant, and appropriate data – and does so ethically (see Principle D). Data collection should not be restricted by favoritism or prejudice. Ethical collection/munging/wrangling follows directly from ethical design and planning. Even if workplace structures mean that the individual who ends up doing the wrangling or munging of data that were collected previously/by someone else's design, the ethical practitioner ensures that their work on that data will not be influenced adversely, and that their work will not exploit vulnerable populations, or perpetuate unfair outcomes.

Principle B. Integrity of data and methods

The ethical statistical practitioner seeks to understand and mitigate known or suspected limitations, defects, or biases in the data or methods and communicates potential impacts on the interpretation, conclusions, recommendations, decisions, or other results of statistical practices.

The ethical statistical practitioner:

B1 Communicates data sources and fitness for use, including data generation and collection processes and known biases. Discloses and manages any conflicts of interest relating to the data sources. Communicates data processing and transformation procedures, including missing data handling.

B2 Is transparent about assumptions made in the execution and interpretation of statistical practices including methods used, limitations, possible sources of error, and algorithmic biases. Conveys results or applications of statistical practices in ways that are honest and meaningful.

B4 Meets obligations to share the data used in the statistical practices, for example, for peer review and replication, as allowable. Respects expectations of data contributors when using or sharing data. Exercises due caution to protect proprietary and confidential data, including all data that might inappropriately harm data subjects.

B6 For models and algorithms designed to inform or implement decisions repeatedly, develops and/or implements plans to validate assumptions and assess performance over time, as needed. Considers criteria and mitigation plans for model or algorithm failure and retirement.

B7 Explores and describes the effect of variation in human characteristics and groups on statistical practice when feasible and relevant.

Since it is focused on the integrity of data and methods around data, five elements of Principle B, as well as the Principle itself, relate directly or indirectly to data collection, wrangling, and/or munging. Identifying fitness for use, and known or suspected defects/limitations or biases in the data will affect later use of that data; when multiple data sets are obtained or munged together, the origins and limitations of all data sets collected/ used must be identified. Data that are not relevant may interfere with "objective and valid interpretation of results". Even if diverse datasets are merged/munged or wrangled in a technically correct way, failures to understand data sources and limitations or biases will undermine the objectives of any project. The influence of errors in data can propagate throughout analyses and affect interpretability/ interpretations in unpredictable ways, so identifying possible sources of error is an important part of both planning and the collection or identification of data to use. The ethical practitioner bears some responsibility for the integrity of data and methods even if they are not the individual who designed the study/project. Moreover, if a collection, munging or wrangling program is created, it (like models and algorithms) needs to be assessed over time, to ensure the processes are ethical and do not become biased (which may lead to violations of A3, exploitation of vulnerable populations or creation/ perpetuation of unfair outcomes).

Principle C. Responsibilities to stakeholders

Those who fund, contribute to, use, or are affected by statistical practices are considered stakeholders. The ethical statistical practitioner respects the interests of stakeholders while practicing in compliance with these Guidelines.

The ethical statistical practitioner:

C1 Seeks to establish what stakeholders hope to obtain from any specific project. Strives to obtain sufficient subject-matter knowledge to conduct meaningful and relevant statistical practice.

C4 Informs stakeholders of the potential limitations on use and re-use of statistical practices in different contexts and offers guidance and alternatives, where appropriate, about scope, cost, and precision considerations that affect the utility of the statistical practice.

C5 Explains any expected adverse consequences from failing to follow through on an agreed-upon sampling or analytic plan.

C7 Understands and conforms to confidentiality requirements for data collection, release, and dissemination and any restrictions on its use established by the data provider (to the extent legally required).

Protects the use and disclosure of data accordingly. Safeguards privileged information of the employer, client, or funder.

Data collection options may vary widely and have dramatic differences in terms of cost (effort and time). Free and plentiful data may be offset by weaker or less precise statistical approaches, while expensive and relatively rarer data may lead to valid and reproducible results. These tradeoffs are an essential part of the ethical practitioner's support of decision making by the funder and/or client. Moreover, the generalizability of results from a highly parameterized data set may be highly limited, weakening the validity of inferences from that data for the scientific community or public. Thus, in the collection, wrangling, and/or munging of data, the ethical practitioner has potentially conflicting responsibilities to these diverse stakeholders. Articulation of decision making by the ethical practitioner, and how these stakeholders' perspectives are balanced, may appeal to a scientific methodology (rather than an appeal to a stakeholder that may be biased). In all cases, restrictions and confidentiality requirements of any data component must be respected.

Principle D. Responsibilities to research subjects, data subjects, or those directly affected by statistical practices

The ethical statistical practitioner does not misuse or condone the misuse of data. They protect and respect the rights and interests of human and animal subjects. These responsibilities extend to those who will be directly affected by statistical practices.

The ethical statistical practitioner:

D1 Keeps informed about and adheres to applicable rules, approvals, and guidelines for the protection and welfare of human and animal subjects. Knows when work requires ethical review and oversight.

D2 Makes informed recommendations for sample size and statistical practice methodology in order to avoid the use of excessive or inadequate numbers of subjects and excessive risk to subjects

D3 For animal studies, seeks to leverage statistical practice to reduce the number of animals used, refine experiments to increase the humane treatment of animals, and replace animal use where possible.

D4 Protects people's privacy and the confidentiality of data concerning them, whether obtained from the individuals directly, other persons, or existing records. Knows and adheres to applicable rules, consents, and guidelines to protect private information.

D5   Uses data only as permitted by data subjects' consent when applicable or considering their interests and welfare when consent is not required. This includes primary and secondary uses, use of repurposed data, sharing data, and linking data with additional data sets.

D6. Considers the impact of statistical practice on society, groups, and individuals. Recognizes that statistical practice could adversely affect groups or the public perception of groups, including marginalized groups. Considers approaches to minimize negative impacts in applications or in framing results in reporting.

D7   Refrains from collecting or using more data than is necessary. Uses confidential information only when permitted and only to the extent necessary. Seeks to minimize the risk of re-identification when sharing de-identified data or results where there is an expectation of confidentiality. Explains any impact of de-identification on accuracy of results.

D8   To maximize contributions of data subjects, considers how best to use available data sources for exploration, training, testing, validation, or replication as needed for the application. The ethical statistical practitioner appropriately discloses how the data is used for these purposes and any limitations.

D9   Knows the legal limitations on privacy and confidentiality assurances and does not over-promise or assume legal privacy and confidentiality protections where they may not apply.

D10 Understands the provenance of the data, including origins, revisions, and any restrictions on usage, and fitness for use prior to conducting statistical practices.

D11 Does not conduct statistical practice that could reasonably be interpreted by subjects as sanctioning a violation of their rights. Seeks to use statistical practices to promote the just and impartial treatment of all individuals.

Principle D and all 11 elements of it quite clearly relate specifically to the collection and wrangling/munging of data. Elements D4, D5, and D6 in particular should protect unknowing data providers from unethical sourcing of data for any project. Any use of data that could create the impression of violating the rights of the data provider is unethical – it is particularly important to guard against this when such a source is to be mixed in with other, possibly ethically-sourced, data. The ethical practitioner strives to offer alternate - ethically sourced – data, and by refusing to engage with/munge or wrangle data with potentially unknown or unethical provenance, s/he acts to discourage others from using such unethically sourced data. However, even ethically sourced data may lead to analyses or summaries that may stigmatize, stereotype, or otherwise represent disproportionate harm to vulnerable groups. Ethically sourced data may still lead to unintended and harmful consequences (violating A3 as well as D11). Understanding the uses to which data are to be put should influence how they are collected, and what is munged together. The ethical practitioner seeks to ensure that data are collected and munged/wrangled ethically and that these efforts are for ethical purposes.

Principle E. Responsibilities to members of multidisciplinary teams

The ethical statistical practitioner:

> E4 Avoids compromising validity for expediency. Regardless of pressure on or within the team, does not use inappropriate statistical practices.

Ethical project planning/study design should lead to collection or identification of data that are ethically sourced, whether the team is engaged in research or non-science work. The ethical practitioner communicates clearly with team colleagues and peers, particularly when there are potential ethical difficulties in the collection or wrangling/munging of data. If some colleagues seek to collect or utilize/munge data that were not ethically sourced because that is more expedient or less expensive than ethically sourced data, this must be resisted to the extent possible. Responsibilities to the data contributors and those affected by statistical practices (Principle D) and those relating to the data (Principle B) preclude the ethical practitioner from participating in projects that involve unethically-sourced, or indefensibly obtained, data – even if this is more expedient, and/or an objective that arises due to pressure on or within a team.

Principle F. Responsibilities to fellow statistical practitioners and the profession

The ethical statistical practitioner:

> F4 Promotes reproducibility and replication, whether results are "significant" or not, by sharing data, methods, and documentation to the extent possible.
>
> F5 Serves as an ambassador for statistical practice by promoting thoughtful choices about data acquisition, analytic procedures, and

data structures among non-practitioners and students. Instills appreciation for the concepts and methods of statistical practice.

Consideration of how others in the profession would consider data that were collected/munged/wrangled can be informative about whether the other relevant GL Principles have been met. If the practitioner fears that their involvement in a data project would reflect badly on them -because data were not source ethically, or, that another statistician would feel uncomfortable using the same data, it suggests that responsibilities for the integrity of the data (B), and those to the data providers (D), have not been met. This is a way to confirm that due consideration of the data collection has been given – even if it is not possible or feasible to actually share the data in question. Unethically sourced data would not be shareable, and data collected under a weak or inappropriate study or sampling design might be sharable but would not be useful to other practitioners. Thus, consideration of F4 and F5 – even if the data cannot actually be shared – can be helpful to ensure the practitioner has met other responsibilities relating to data collection, munging and wrangling.

Principle G. Responsibilities of leaders, supervisors, and mentors in statistical practice

G1 Ensure appropriate statistical practice that is consistent with these Guidelines. Protect the statistical practitioners who comply with these Guidelines, and advocate for a culture that supports ethical statistical practice.

G5 Establish a culture that values validation of assumptions, and assessment of model/algorithm performance over time and across relevant subgroups, as needed. Communicate with relevant stakeholders regarding model or algorithm maintenance, failure, or actual or proposed modifications.

The ethical practitioner recognizes that ethical practice should be explicitly supported in the workplace; leaders, supervisor, and mentors must recognize when behaviors are contrary to the GLs and should therefore not be allowed/followed. This is particularly important when data are not ethically collected, or when unethically sourced data are required to be introduced into otherwise ethically sourced data. The ethical practitioner directly supports the freedom and responsibility of statistical practitioners who comply with these GLs and strengthens the profession by refusing, and supporting others who also refuse, to compromise data contributor's rights, or otherwise undermine the rights of individuals with respect to their own data.

Principle H. Responsibilities regarding potential misconduct

The ethical statistical practitioner:

> H2 Avoids condoning or appearing to condone statistical, scientific, or professional misconduct. Encourages other practitioners to avoid misconduct or the appearance of misconduct.

The ethical practitioner engages in, and directly and indirectly supports, the collection and use of only ethically sourced data. Allowing unethically-sourced data to contaminate otherwise-ethical data undermines the legitimacy of the entire project – and so condones the unethical sourcing of at least some data. Following Principle D (Responsibilities to research subjects, data subjects, or those directly affected by statistical practices) – and striving to ensure that others do so as well – cannot be interpreted as supporting or condoning incompetent and/or unprofessional behavior. When plentiful, but unethically sourced, data are sought or utilized, it directly prioritizes expedience over validity (contrary to E4) as well as failing in Responsibilities to research subjects, data subjects, or those directly affected by statistical practices (Principle D), constituting both statistical misconduct and detrimental practice.

ACM Code of Ethics on: data collection/munging/wrangling

Overall: The Preamble states that the Code is designed to inspire and guide the ethical conduct of "anyone who uses computing technology in an impactful way…based on the understanding that the public good is always the primary consideration.

The ethical computing professional, encompassing all who use computing technology in a way that can impact others, which must include the collection of data, should keep the public good, and the minimization of harms, as their primary focus. If data are easy to collect, but doing so may undermine the public good or disenfranchise any individual or group of individuals, the ethical professional does not collect such data.

1. General Ethical Principles

> 1.1 This principle, which concerns the quality of life of all people, affirms an obligation of computing professionals, both individually and collectively, to use their skills for the benefit of society, its members, and the environment surrounding them. This obligation includes promoting fundamental human rights and protecting each individual's right to autonomy. An essential aim of computing

professionals is to minimize negative consequences of computing, including threats to health, safety, personal security, and privacy. When the interests of multiple groups conflict, the needs of those less advantaged should be given increased attention and priority.

1.2 To minimize the possibility of indirectly or unintentionally harming others, computing professionals should follow generally accepted best practices unless there is a compelling ethical reason to do otherwise. Additionally, the consequences of data aggregation and emergent properties of systems should be carefully analyzed.

1.3 A computing professional should be transparent and provide full disclosure of all pertinent system capabilities, limitations, and potential problems to the appropriate parties.

1.4 The use of information and technology may cause new, or enhance existing, inequities. Technologies and practices should be as inclusive and accessible as possible and computing professionals should take action to avoid creating systems or technologies that disenfranchise or oppress people.

1.6 Computing professionals should only use personal information for legitimate ends and without violating the rights of individuals and groups. This requires taking precautions to prevent re- identification of anonymized data or unauthorized data collection, ensuring the accuracy of data, understanding the provenance of the data, and protecting it from unauthorized access and accidental disclosure.

Computing professionals should establish transparent policies and procedures that allow individuals to understand what data is being collected and how it is being used, to give informed consent for automatic data collection, and to review, obtain, correct inaccuracies in, and delete their personal data.

Only the minimum amount of personal information necessary should be collected in a system. The retention and disposal periods for that information should be clearly defined, enforced, and communicated to data subjects. Personal information gathered for a specific purpose should not be used for other purposes without the person's consent. Merged data collections can compromise privacy features present in the original collections. Therefore, computing professionals should take special care for privacy when merging data collections.

The ethical computing professional collects only the minimum amount of personal information necessary to accomplish whatever the specific task of the system is known to be. Collecting unlimited personal information because a system can, without notifying individuals whose data is being collected of the purposes for which the data are being collected, directly violates Principles 1.3 and 1.6, and may also result in bias, decrement of the public trust in computing professionals, and misuse of data. The ethical practitioner should transparently report to data contributors exactly what data is being collected, otherwise, consent for data collection cannot be characterized as "informed". This may result in less data being collected – and will also ensure that the universally-accepted best practice of informed consent for data collection (i.e., avoiding "unauthorized data collection") is adhered to. Collecting less data may create a net negative for the employer or client while avoiding harms to the public and the public's rights to autonomy, and their right to determine whether and to what extent their data are collected. As stated in Principle 1.1, "When the interests of multiple groups conflict, the needs of those less advantaged should be given increased attention and priority." The individual should be considered "less advantaged" as compared to the funder/client, because any harms that accrue because of unauthorized data collection necessarily accrue to the individual (and more generally, to the public).

2. Professional Responsibilities

2.1   The dignity of employers, employees, colleagues, clients, users, and anyone else affected either directly or indirectly by the work should be respected throughout the process. Computing professionals should respect the right of those involved to transparent communication about the project. Professionals should be cognizant of any serious negative consequences affecting any stakeholder that may result from poor quality work and should resist inducements to neglect this responsibility.

2.2   Professional competence starts with technical knowledge and with awareness of the social context in which their work may be deployed.

2.3   Rules that are judged unethical should be challenged. A rule may be unethical when it has an inadequate moral basis or causes recognizable harm.

2.4   Whenever appropriate, computing professionals should seek and utilize peer and stakeholder review.

2.5   Extraordinary care should be taken to identify and mitigate potential risks in machine learning systems. A system for which future risks cannot be reliably predicted requires frequent reassessment of risk as the system evolves in use, or it should not be deployed.

2.8   Individuals and organizations have the right to restrict access to their systems and data so long as the restrictions are consistent with other principles in the Code. Consequently, computing professionals should not access another's computer system, software, or data without a reasonable belief that such an action would be authorized or a compelling belief that it is consistent with the public good. A system being publicly accessible is not sufficient grounds on its own to imply authorization.

2.9 Computing professionals should perform due diligence to ensure the system functions as intended, and take appropriate action to secure resources against accidental and intentional misuse, modification, and denial of service. In cases where misuse or harm are predictable or unavoidable, the best option may be to not implement the system.

The ethical computing professional does not access or collect an individual's personal data without authorization or the belief that this access would be authorized. When a system can be predicted to be misused (to collect data without authorization and knowledge of the individual or organization whose data is to be collected), the best option for the ethical practitioner may be "not to implement the system". As specified in 2.6, "A computing professional's ethical judgment should be the final guide in deciding whether to work on the assignment." Even if (when) another (not ethical) practitioner may implement a system that collects data in unauthorized ways, the ethical practitioner does not create or implement such a system.

3. Professional Leadership Principles

3.1 People—including users, customers, colleagues, and others affected directly or indirectly— should always be the central concern in computing. The public good should always be an explicit consideration when evaluating tasks associated with research, requirements analysis, design, implementation, testing, validation, deployment, maintenance, retirement, and disposal. Computing professionals should keep this focus no matter which methodologies or techniques they use in their practice.

3.4 Leaders should pursue clearly defined organizational policies that are consistent with the Code and effectively communicate them to relevant stakeholders. In addition, leaders should encourage and reward compliance with those policies, and take appropriate action when policies are violated. Designing or implementing processes that deliberately or negligently violate, or tend to enable the violation of, the Code's principles is ethically unacceptable.

3.5 Computing professionals should be fully aware of the dangers of oversimplified approaches, the improbability of anticipating every possible operating condition, the inevitability of software errors, the interactions of systems and their contexts, and other issues related to the complexity of their profession—and thus be confident in taking on responsibilities for the work that they do.

The ethical computing professional keeps those who are affected "directly or indirectly" by computing as their central concern: when a business may benefit from collecting data without explicit and informed consent, but the individuals from whom data are collected without consent (or their rights to autonomy, and to be informed about the uses to which their data are collected and put) may be harmed or violated, that directly violates multiple ACM Code Principles. The collection of data must follow globally recognized rights of the individual. Wrangling or munging data must follow from informed consent for the purposes to which data will be put. Provenance of data must be established before wrangling and munging, otherwise the computing system (and its developers and deployers) violate the ACM Code of Ethics. The stakeholder analysis (SHA) table below shows some harms, benefits, unknown outcomes, and unknowable outcomes, relating to following the GLs and CE in data collection/munging/wrangling.

Potential result: Stakeholder:	HARM	BENEFIT	UNKNOWN	UNKNOWABLE
YOU	Takes time, adds effort (to find acceptable data)	Informed consent ensures compliance with CE/GL and respect for autonomy	Informed consent protects against violations of data contributors' rights	Supports autonomy but may create bias as groups differentially opt in/out
Your boss/client	Takes time & may limit data collected, thus generalizability of results	Informed consent ensures compliance and respect for autonomy	Differential consent may affect munging (mungeability)	Differential consent may lead to bias in results
Unknown individuals	Uninformed, unconsented, or unauthorized data collection violates rights of autonomy and best practices in data collection.	Mitigates risk of bias/privacy breach/harms	Differential consent may lead to bias in results	
Employer	Takes time & may limit data collected, thus generalizability of results	Informed consent ensures compliance with CE/GL and respect for autonomy	Differential consent may lead to bias in results	Could limit business/profit
Colleagues	May complicate colleagues' work (challenging them to follow GL/CE)	Supports compliance and respect for autonomy across all colleagues	Differential consent may lead to bias in results that affect upstream systems	Commitment to informed consent can enhance colleagues' trust in ethically sourced data

Potential result: Stakeholder:	HARM	BENEFIT	UNKNOWN	UNKNOWABLE
Profession	Oversimplified approaches may lead to collected data, but violate individual rights, undermining the profession	Informed consent ensures compliance and respect for autonomy and challenges practitioners to respect autonomy		Commitment to GL/CE strengthens trust in profession and enables solutions/ Conversations
Public/public trust	Unauthorized data collection can undermine public trust in systems and the profession	Transparency and explicit respect for autonomy supports public trust in the profession	Transparency supports trust, even if misuse does occur; policies supporting autonomy and informed consent can tend to inform the public.	Engagement with GL/CE promotes trust even if misuse does occur

Table 2.3.1. *Stakeholder Analysis template: data collection/munging/wrangling*

Questions for Discussion:

Discuss similarities and differences between the ASA GLs and ACM CE principles that are discussed above, in terms of how they apply to data collection, wrangling, and/or munging.

What are the most striking similarities (e.g., what is unexpectedly similar)? Are there particular differences that stand out because they are A) highly predictable? B) unexpected? C) have different benefits or harms to different stakeholders?

Rank GL/CE Principles from most to least important to data collection, munging, and wrangling, using their own language or your own perspective (and identify which you chose in your answer). What does the rankability mean for you or for the discipline of statistics and data science?

Can following the GL/CE affect how you collect data, wrangle, or munge data in your typical work? How/How not?

Is the SHA in Table 2.3.1 enough to ensure ethical collection, wrangling, or munging of data in the typical workplace or project? Why/why not?

Discuss how the ASA and ACM are, or are not, fully aligned with current "policies regarding human subjects, live vertebrate animal subjects in research, and safe laboratory practices".

Discuss how the ASA and ACM are, or are not, fully aligned with current policies (for your institution, organization, or lab/work team) on "data acquisition and laboratory tools; management, sharing and ownership".

Chapter 2.4
Analysis (perform or program to perform)

NOTE: "analysis" has different meanings for straight computation (ACM) and straight statistical (ASA) users, with blended meanings for those who use computation extensively in their statistical practice (or who use statistics in their computational practice). Specifically, ASA GLs refer to the analysis of data, and in some cases to an analysis of how people use the data or the analysis (evaluation) of others' use of data, results, or methods. By contrast, the ACM CE refers to the analysis of risks, the evaluation (and implied analysis) of systems or their plans/designs, and to the analysis of activities, or the effects of computing on stakeholders.

ASA Ethical Guidelines (GLs) for Statistical Practice: on analysis

Overall: The first paragraph of the preamble to the GLs states that:

"In these Guidelines, "statistical practice" includes activities such as: designing the collection of, summarizing, processing, *analyzing*, interpreting, or presenting, data; as well as model or algorithm development and deployment. Throughout these Guidelines, the term "statistical practitioner" includes *all those who engage in statistical practice, regardless of job title, profession, level, or field of degree."* (Emphasis added)

The Ethical Guidelines for Statistical Practice are intended for "all those who engage in statistical practice, regardless of job title, profession, level, or field of degree." Analysis of data is naturally essential in statistical practice, even for those who do not self-identify as a statistician, or whose role or training in "data analysis" may not have focused on the active role that a programmer plays in creating a program or algorithm to carry out analyses rather than doing the analyses more directly (i.e., using statistical software or programs like R). The ethical practitioner – no matter what their title, nor how much of their time is spent in actual analysis - strives to generate results that are, and supports the design of work that is, transparent, reproducible, and valid. Competent and ethical analysis is absolutely essential in these efforts. The ethical practitioner both acts in a manner consistent with the GLs in their own analyses, and also encourages others to do so in theirs. The conduct of analyses in every project must include considerations of the potential harm that results from that analysis can bring or create. Society "benefits from informed judgments supported by ethical statistical practice", meaning that the analyses we plan and execute are intrinsically linked to those judgments – and thereby, any benefits the analyses do or are intended to create. Even if one individual conducts analyses and others do the planning and reporting, it is the analyses and their results that will form the basis of those judgments. The ethical practitioner is aware of this relationship between their analyses and those judgments and decisions.

More specifically, the ethical practitioner recognizes their obligations to practice statistics ethically, even if that is not their job title or primary function, and whether or not they perceive themselves as doing the analyses or if they create programs or algorithms to do the analyses.

Principle A: Professional integrity and accountability

Professional integrity and accountability require taking responsibility for one's work. Ethical statistical practice supports valid and prudent decision making with appropriate methodology. The ethical statistical practitioner represents their capabilities and activities honestly, and treats others with respect.

The ethical statistical practitioner:

A1 Takes responsibility for evaluating potential tasks, assessing whether they have (or can attain) sufficient competence to execute each task, and that the work and timeline are feasible. Does not solicit or deliver work for which they are not qualified, or that they would not be willing to have peer reviewed.

A2 Uses methodology and data that are valid, relevant, and appropriate, without favoritism or prejudice, and in a manner intended to produce valid, interpretable, and reproducible results.

A3 Does not knowingly conduct statistical practices that exploit vulnerable populations or create or perpetuate unfair outcomes.

A4 Opposes efforts to predetermine or influence the results of statistical practices, and resists pressure to selectively interpret data.

Principle A, as well as these four identified elements, relate specifically to how important it is for the ethical practitioner to carry out or program analyses competently and transparently, using appropriate methodology and not selecting methods in a way that is limited or restricted by favoritism or prejudice. Favoritism and prejudice in the selection of methods, or assumptions made during analyses, may arise from, or create, conflicts of interest. The contributions of others to an analysis plan must be acknowledged, for example, if someone designed or planned the analyses and another analyst executes it. Even if workplace structures mean that the individual who plans or designs a study/project/program will not be the one executing the analysis, the ethical practitioner seeks to ensure that the analyses they program or execute are appropriate and valid, and accepts responsibility for their work.

Principle B. Integrity of data and methods

The ethical statistical practitioner seeks to understand and mitigate known or suspected limitations, defects, or biases in the data or methods and communicates potential impacts on the interpretation, conclusions, recommendations, decisions, or other results of statistical practices.

The ethical statistical practitioner:

B1 Communicates data sources and fitness for use, including data generation and collection processes and known biases. Discloses and manages any conflicts of interest relating to the data sources. Communicates data processing and transformation procedures, including missing data handling.

B2 Is transparent about assumptions made in the execution and interpretation of statistical practices including methods used, limitations, possible sources of error, and algorithmic biases. Conveys results or applications of statistical practices in ways that are honest and meaningful.

B3 Communicates the stated purpose and the intended use of statistical practices. Is transparent regarding a priori versus post hoc objectives and planned versus unplanned statistical practices. Discloses when multiple comparisons are conducted, and any relevant adjustments.

B4 Meets obligations to share the data used in the statistical practices, for example, for peer review and replication, as allowable. Respects expectations of data contributors when using or sharing data. Exercises due caution to protect proprietary and confidential data, including all data that might inappropriately harm data subjects.

B5 Strives to promptly correct substantive errors discovered after publication or implementation. As appropriate, disseminates the correction publicly and/or to others relying on the results.

B6 For models and algorithms designed to inform or implement decisions repeatedly, develops and/or implements plans to validate assumptions and assess performance over time, as needed. Considers criteria and mitigation plans for model or algorithm failure and retirement.

B7 Explores and describes the effect of variation in human characteristics and groups on statistical practice when feasible and relevant.

Every element in Principle B, as well as the Principle itself, relates directly or indirectly to analysis. Awareness of whether there are known or suspected defects/limitations or biases in the data that can have an effect in later use of that data is clearly important for the analysis of that data. Ethical analysis requires that data are well-understood; without this background, even if an analysis is done technically correctly, "objective and valid interpretation of results" may not follow if the analyses were correctly done on corrupt or biased data. Professional, and ethical, analysis plans must all keep these endpoints in mind, and ensuring the integrity of the data that are analyzed is essential. Analysis plans should include contingency plans in case any error is (eventually) discovered. The ethical practitioner bears responsibility for the integrity of data and appropriateness of the analytic methods, because otherwise, technically correct analyses will not support any decisions made on the basis of those results.

Principle C. Responsibilities to stakeholders

Those who fund, contribute to, use, or are affected by statistical practices are considered stakeholders. The ethical statistical practitioner respects the interests of stakeholders while practicing in compliance with these Guidelines.

The ethical statistical practitioner:

C1 Seeks to establish what stakeholders hope to obtain from any specific project. Strives to obtain sufficient subject-matter knowledge to conduct meaningful and relevant statistical practice.

C2 Regardless of personal or institutional interests or external pressures, does not use statistical practices to mislead any stakeholder.

C3 Uses practices appropriate to exploratory and confirmatory phases of a project, differentiating findings from each so the stakeholders can understand and apply the results.

C4 Informs stakeholders of the potential limitations on use and re-use of statistical practices in different contexts and offers guidance and alternatives, where appropriate, about scope, cost, and precision considerations that affect the utility of the statistical practice.

C5 Explains any expected adverse consequences from failing to follow through on an agreed-upon sampling or analytic plan.

C8 Prioritizes both scientific integrity and the principles outlined in these Guidelines when interests are in conflict.

Whether or not the analyst is also the individual who planned or programmed the analysis, the ethical practitioner should be aware of, and prepared for, contingencies that may affect stakeholders, including the wider scientific community, the public, and the funder or employer. As the analyst ensures they are following Principle B, these contingencies may arise (e.g., if potential confounders are identified and affect the analysis results). If the data are found to be unethically sourced (see Principle D), valid alternative analytic methods may be needed. The ethical practitioner has responsibilities to follow an agreed-upon analytic plan, but if such a plan is created without following Principles A and B (e.g., if inappropriate analyses were planned; or if the data are discovered to have defects or biases that will yield irreproducible results), then the ethical practitioner should revisit their responsibilities to the funder/client and consider an alternative plan. Even in non-scientific contexts, scientific integrity, which promotes effective interpretation based on specific hypotheses and appropriate methods, is relevant.

Principle D. Responsibilities to research subjects, data subjects, or those directly affected by statistical practices

The ethical statistical practitioner does not misuse or condone the misuse of data. They protect and respect the rights and interests of human and animal subjects. These responsibilities extend to those who will be directly affected by statistical practices.

The ethical statistical practitioner:

D2 Makes informed recommendations for sample size and statistical practice methodology in order to avoid the use of excessive or inadequate numbers of subjects and excessive risk to subjects

D5 Uses data only as permitted by data subjects' consent when applicable or considering their interests and welfare when consent is not required. This includes primary and secondary uses, use of repurposed data, sharing data, and linking data with additional data sets.

D6 Considers the impact of statistical practice on society, groups, and individuals. Recognizes that statistical practice could adversely affect groups or the public perception of groups, including marginalized groups. Considers approaches to minimize negative impacts in applications or in framing results in reporting.

D8   To maximize contributions of data subjects, considers how best to use available data sources for exploration, training, testing, validation, or replication as needed for the application. The ethical statistical practitioner appropriately discloses how the data is used for these purposes and any limitations.

D10 Understands the provenance of the data, including origins, revisions, and any restrictions on usage, and fitness for use prior to conducting statistical practices.

D11 Does not conduct statistical practice that could reasonably be interpreted by subjects as sanctioning a violation of their rights. Seeks to use statistical practices to promote the just and impartial treatment of all individuals.

Even when an analysis plan is provided to the analyst, and a data set is provided, the ethical practitioner accepts responsibility for ethical analysis. Specifically, elements D5 and D6 specify that the ethical practitioner does not execute an analysis of data lacking appropriate approvals, nor when the analysis might create the impression of violating (or actually violate) the rights of those whose data are to be used. The primary obligation is for the ethical practitioner to refuse to execute an analysis plan using unethically sourced data, or data that may lead to violations of data contributors' rights. If possible, s/he should also try to discourage others from using such unethically sourced data. Not only are analyses of ethically sourced data more reproducible and valid, they also concretely represent the respect of the ethical practitioner for the animals and humans whose data are analyzed.

Principle E. Responsibilities to members of multidisciplinary teams

The ethical statistical practitioner:

E4 Avoids compromising validity for expediency. Regardless of pressure on or within the team, does not use inappropriate statistical practices.

In cases where the person analyzing the data is one of a team, whether or not the team is engaged in research, the ethical practitioner has responsibilities to follow these GLs. It is important for the analyst to recognize that other team members may not be familiar with, or feel obliged to follow, these GLs (even if they are also statistical practitioners), and that does not affect the analyst's obligations to do so. Maintaining a professional perspective in the face of team pressures can be very challenging, and these GLs are intended to support the practitioner who seeks to do their work ethically.

Whether the analyst was also able to ensure that the analysis plan was ethically designed, and intended to be applied to ethically sourced data, or if a plan plus some data were simply handed off, the ethical practitioner ensures that *what will be reported* was ethically designed and executed. The ethical practitioner does not simply execute an analysis plan they were handed, without considering the source of the data or the appropriateness of the methods planned.

Principle F. Responsibilities to fellow statistical practitioners and the Profession

Statistical practices occur in a wide range of contexts. Irrespective of job title and training, those who practice statistics have a responsibility to treat statistical practitioners, and the profession, with respect. Responsibilities to other practitioners and the profession include honest communication and engagement that can strengthen the work of others and the profession.

The ethical statistical practitioner:

F4 Promotes reproducibility and replication, whether results are "significant" or not, by sharing data, methods, and documentation to the extent possible.

F5 Serves as an ambassador for statistical practice by promoting thoughtful choices about data acquisition, analytic procedures, and data structures among non-practitioners and students. Instills appreciation for the concepts and methods of statistical practice.

Ethical analysis follows from plans, and uses data, that are ethically sourced and derived. However, there may be a range of approaches that can be used in any given analytic context, and while the ethical statistical practitioner has responsibilities to ensure that analyses and results are appropriate methodologically, and competently executed, alternative approaches may also be appropriate. In the efforts to ensure that Principles A-E are met, the ethical statistical practitioner is respectful of other statisticians, focusing on the science, methodologies, and the reproducibility of interpretable results. When analytic methodological perspectives are in conflict, thoughtful synthesis of diverse results will strengthen the discipline of statistics as well as supporting a better understanding of the interpretability and reproducibility of results.

Principle G. Responsibilities of leaders, supervisors, and mentors in statistical practice

Those leading, supervising, or mentoring statistical practitioners are expected to:

G1 Ensure appropriate statistical practice that is consistent with these Guidelines. Protect the statistical practitioners who comply with these Guidelines, and advocate for a culture that supports ethical statistical practice.

G5 Establish a culture that values validation of assumptions, and assessment of model/algorithm performance over time and across relevant subgroups, as needed. Communicate with relevant stakeholders regarding model or algorithm maintenance, failure, or actual or proposed modifications.

As noted, the individual who is tasked with executing an analysis plan prepared by someone else, on data they were not involved in collecting, sometimes has a uniquely difficult burden in following the ASA GLs. Leaders, supervisors, and mentors should support the analyst's efforts to ensure that the plan is appropriate, and the data are ethically sourced, prior to executing an analysis plan. Complete and transparent documentation of planning and design by the ethical statistical practitioner will enable the analyst (if that is a different person) to see that the GLs were followed; but following G1 means that sound plans should be executed on ethically sourced data, and unsound plans should not be implemented, nor should unethically sourced data be analyzed. The ethical statistical practitioner contributes to an ethical workplace in leadership role by following G1, G5 (and all the GLs).

Principle H. Responsibilities regarding potential misconduct

The ethical statistical practitioner:

H2 Avoids condoning or appearing to condone statistical, scientific, or professional misconduct. Encourages other practitioners to avoid misconduct or the appearance of misconduct.

The ethical practitioner engages in, and directly and indirectly supports, ethical analysis so as to avoid condoning (or engaging in) incompetent and unprofessional behavior. While not illegal, choosing a simple(r) analytic method that can be technically correct but contextually inappropriate is an instance of prioritizing expedience over validity (contrary to E3), and is an example of statistical and scientific misconduct. A simpler analytic method that does not lead to different inferences or interpretations can be utilized (and the results should be obtained and compared to support these choices). Differences of opinion on analytic methodology, and honest errors in application of those methods, do not constitute misconduct. In a workplace where ethical statistical practice is supported (see Principle G), conversations/discussions about choices should be common; transparency and completeness in reporting will also strengthen the perception of ethical and competent alternative analytic approaches.

ACM Code of Ethics: on Analysis

Overall: The Preamble states:

When thinking through a particular issue, a computing professional may find that multiple principles should be taken into account, and that different principles will have different relevance to the issue. Questions related to these kinds of issues can best be answered by thoughtful consideration of the fundamental ethical principles, understanding that the public good is the paramount consideration.

The ethical computing professional has a responsibility to analyze the work they do/are asked to do to ensure that decisions are justifiable and ethical to promote the public good and minimize harms and risks of harms.

1. General Ethical Principles

1.1 This principle, which concerns the quality of life of all people, affirms an obligation of computing professionals, both individually and collectively, to use their skills for the benefit of society, its members, and the environment surrounding them. This obligation includes promoting fundamental human rights and protecting each individual's right to autonomy. An essential aim of computing professionals is to minimize negative consequences of computing, including threats to health, safety, personal security, and privacy. When the interests of multiple groups conflict, the needs of those less advantaged should be given increased attention and priority.

Computing professionals should consider whether the results of their efforts will respect diversity, will be used in socially responsible ways, will meet social needs, and will be broadly accessible.

1.2 Avoiding harm begins with careful consideration of potential impacts on all those affected by decisions. When harm is an intentional part of the system, those responsible are obligated to ensure that the harm is ethically justified. In either case, ensure that all harm is minimized.

To minimize the possibility of indirectly or unintentionally harming others, computing professionals should follow generally accepted best practices unless there is a compelling ethical reason to do otherwise. Additionally, the consequences of data aggregation and emergent properties of systems should be carefully analyzed.

A computing professional has an additional obligation to report any signs of system risks that might result in harm. If leaders do not act to curtail or mitigate such risks, it may be necessary to "blow the whistle" to reduce potential harm. However, capricious or misguided reporting of risks can itself be harmful. Before reporting risks, a computing professional should carefully assess relevant aspects of the situation.

1.3 A computing professional should be transparent and provide full disclosure of all pertinent system capabilities, limitations, and potential problems to the appropriate parties.

1.4 The use of information and technology may cause new, or enhance existing, inequities. Technologies and practices should be as inclusive and accessible as possible and computing professionals should take action to avoid creating systems or technologies that disenfranchise or oppress people.

1.6 Computing professionals should only use personal information for legitimate ends and without violating the rights of individuals and groups. This requires taking precautions to prevent re- identification of anonymized data or unauthorized data collection, ensuring the accuracy of data, understanding the provenance of the data, and protecting it from unauthorized access and accidental disclosure.\ Computing professionals should establish transparent policies and procedures that allow individuals to understand what data is being collected and how it is being used, to give informed consent for automatic data collection, and to review, obtain, correct inaccuracies in, and delete their personal data.

Only the minimum amount of personal information necessary should be collected in a system. The retention and disposal periods for that information should be clearly defined, enforced, and communicated to data subjects. Personal information gathered for a specific purpose should not be used for other purposes without the person's consent. Merged data collections can compromise privacy features present in the original collections. Therefore, computing professionals should take special care for privacy when merging data collections.

As noted at the beginning of this chapter, the ACM CE is naturally not focused on the same type of "analysis" that the ASA GLs is focused on. However, much of ACM Principle 1 entails or follows from an analysis of systems or behaviors. ACM 1.6 relates perhaps most explicitly to the type of analysis that the ASA GLs do: ensuring that data to be utilized or analyzed has been ethically sourced, based on known provenance, is an essential part of both ASA and ACM perspectives on data analysis. Other elements of ACM Principle 1 relate to the analysis of systems, or the context in which systems will be developed or deployed. While ASA GL readers are also expected to base ethical decisions on their "analysis" of the context in which they work, that is only secondary to the emphasis on utilizing the correct method of data (inferential) analysis. The remainder of relevant parts of ACM Principle 1 for "analysis" focus on the ACM member's/ethical computing professional's attention to details relating to systems and their effects on stakeholders (particularly, the public).

An implicit type of analysis is that which supports the ethical practitioner transparently reporting to data contributors exactly what data is being collected, otherwise, consent for data collection cannot be characterized as "informed". The computing professional must understand fully what the data collection system is doing, and for what purposes – as well as how these are presented to the potential data contributor. That represents a complex level of analysis but is essential for all of the ACM CE Principles relating to ethical data collection, security, maintenance, and ultimately, disposal.

Yet another analysis that the ACM CE implies is essential to ethical practice is stated in Principle 1.1, "When the interests of multiple groups conflict, the needs of those less advantaged should be given increased attention and priority." This represents an analysis similar to the stakeholder analysis we have discussed – since the ACM articulates a specific responsibility by all ethical practitioners to protect individual rights and the public good, ACM Principle 1.1 (among others) points to an obligation to analyze how harms or benefits may (or do) accrue to the individual (and more generally, to the public).

2. Professional Responsibilities

2.1 The dignity of employers, employees, colleagues, clients, users, and anyone else affected either directly or indirectly by the work should be respected throughout the process. Computing professionals should respect the right of those involved to transparent communication about the project. Professionals should be cognizant of any serious negative consequences affecting any stakeholder that may result from poor quality work and should resist inducements to neglect this responsibility.

2.2 Professional competence starts with technical knowledge and with awareness of the social context in which their work may be deployed.

2.3 Rules that are judged unethical should be challenged. A rule may be unethical when it has an inadequate moral basis or causes recognizable harm.

2.4 Computing professionals should also provide constructive, critical reviews of others' work.

2.5 Computing professionals are in a position of trust, and therefore have a special responsibility to provide objective, credible evaluations and testimony to employers, employees, clients, users, and the public. Computing professionals should strive to be perceptive, thorough, and objective when evaluating, recommending, and presenting system descriptions and alternatives. Extraordinary care should be taken to identify and mitigate potential risks in machine learning systems. A system for which future risks cannot be reliably predicted requires frequent reassessment of risk as the system evolves in use, or it should not be deployed. Any issues that might result in major risk must be reported to appropriate parties.

2.6 A computing professional is responsible for evaluating potential work assignments. This includes evaluating the work's feasibility and advisability, and making a judgment about whether the work assignment is within the professional's areas of competence. If at any time before or during the work assignment the professional identifies a lack of a necessary expertise, they must disclose this to the employer or client.

2.8 Under exceptional circumstances a computing professional may use unauthorized access to disrupt or inhibit the functioning of malicious systems; extraordinary precautions must be taken in these instances to avoid harm to others.

2.9 Computing professionals should perform due diligence to ensure the system functions as intended, and take appropriate action to secure resources against accidental and intentional misuse, modification, and denial of service. As threats can arise and change after a system is deployed, computing professionals should integrate mitigation techniques and policies, such as monitoring, patching, and vulnerability reporting.

In cases where misuse or harm are predictable or unavoidable, the best option may be to not implement the system.

The professional responsibilities of the ethical computing professional involve extensive contextual and stakeholder analysis, whether to determine whether rules are ethical and should be followed, or are not ethical/should not be followed (2.3), to provide constructive feedback (2.4), or to determine that the risk for misuse or harm are so great that the system should not be implemented. The ACM CE is clear in its perspective that the public good, and individual rights, are to be prioritized; however, to determine what serves these priorities best, careful and thoughtful analysis is required.

3. Professional Leadership Principles

3.1 People—including users, customers, colleagues, and others affected directly or indirectly— should always be the central concern in computing. The public good should always be an explicit consideration when evaluating tasks associated with research, requirements analysis, design, implementation, testing, validation, deployment, maintenance, retirement, and disposal. Computing professionals should keep this focus no matter which methodologies or techniques they use in their practice.

3.4 Leaders should pursue clearly defined organizational policies that are consistent with the Code and effectively communicate them to relevant stakeholders. In addition, leaders should encourage and reward compliance with those policies, and take appropriate action when policies are violated. Designing or implementing processes that deliberately or negligently violate, or tend to enable the violation of, the Code's principles is ethically unacceptable.

3.5 Computing professionals should be fully aware of the dangers of oversimplified approaches, the improbability of anticipating every possible operating condition, the inevitability of software errors, the interactions of systems and their contexts, and other issues related to the complexity of their profession—and thus be confident in taking on responsibilities for the work that they do.

Reiterating the emphasis on the public good and avoiding harm, the ethical computing professional as well as those in leadership roles must consider (analyze) stakeholder benefits and harms, as well as contextual cues (or rules/policies). Analysis such as the stakeholder analysis we have been utilizing can help the ethical practitioner identify both those who are affected "directly or indirectly" by computing and also whether or not harms or benefits accrue to them.

4. Compliance with the Code

4.1 Computing professionals who recognize breaches of the Code should take actions to resolve the ethical issues they recognize, including, when reasonable, expressing their concern to the person or persons thought to be violating the Code.

4.2 ACM members who recognize a breach of the Code should consider reporting the violation.

As discussed for other ACM Principles, the identification of failures to comply with the ACM CE will require analysis, particularly of the harms and benefits that accrue to the various stakeholders. The stakeholder analysis (SHA) table below shows some harms, benefits, unknown outcomes, and unknowable outcomes, relating to following the GLs and CE in analysis.

Potential result: Stakeholder:	HARM	BENEFIT	UNKNOWN	UNKNOWABLE
YOU	Takes time, adds effort; may identify problems that require additional time/effort	Thoughtful analysis ensures compliance with CE/GL, appropriate work, and identification of risks/harms	Competent analysis (ASA) supports reproducible results; effective analysis (ACM) promotes public good. Both take time.	All harms cannot be identified; typically no "one right" answer in a statistical analysis.
Your boss/client	Takes time, adds effort; may identify problems that require additional time/effort or that undermine project totally	Analysis ensures compliance with CE/GL, appropriate work (ASA) and identification of risks/harms (ACM)	Differential attention to analysis (ASA and ACM) may unbalance teams or overall project.	Practitioner's competence is supported by careful analysis (ASA and ACM) but may identify problems/not yield expected results.
Unknown individuals	Analysis required or ACM CE violated; weak statistical analysis undermines reproducibility and rigor.	Identifies and allows mitigation of risks of harms (ACM); supports competent inferences (ASA)		
Employer	Takes time, adds effort; may identify problems that require additional time/effort or that undermine project totally	Analysis ensures compliance with CE/GL, appropriate work (ASA) and identification of risks/harms (ACM)	Identification of risks/harms (ACM) may require resources to fix/address. Appropriate analysis (ASA) may not yield desired results.	Could limit business/profit (ASA or ACM), or demonstrate a project to be un-deployable (ACM).

Potential result: Stakeholder:	HARM	BENEFIT	UNKNOWN	UNKNOWABLE
Colleagues	May complicate colleagues' work if analysis identifies problems.	Supports compliance and excellence across all colleagues	ACM/ASA analysis may each yield results that affect upstream systems	Project could be delayed until risks of harms/harms are addressed
Profession	Focus on each type of ACM CE analysis may undermine agility and responsivity in the profession; extensive sensitivity analysis (ASA) may lead to frustration without resolution of scientific questions.	Focus on competent analysis ensures compliance with CE/GL – strengthening these – while appropriate work (ASA) and identification of risks/harms (ACM) increase trust in profession.		Commitment to GL/CE can strengthen trust in profession and can enable innovation as well as conversations about ethical practice.
Public/public trust	Incomplete analysis may fail to identify harms/risks, or may lead to false positive/irreproducible results, undermining public trust in systems and the profession.	Transparency and explicit respect for professional standards (ASA, ACM) supports public trust in the profession.	Transparency supports trust, even if risk/harms are identified; can tend to inform the public about limitations inherent in the profession.	

Table 2.4.1. *Stakeholder Analysis template: Analysis*

Questions for Discussion:

Discuss similarities and differences between the ASA GLs and ACM CE principles that are discussed above, in terms of how they apply to either statistical (ASA) or other (ACM) analysis.

What are the most striking similarities (e.g., what is unexpectedly similar)? Are there particular differences that stand out because they are A) highly predictable? B) unexpected? C) have different benefits or harms to different stakeholders?

Rank GL/CE elements from most to least important to analysis. What does the rankability mean for you or for the discipline of statistics and data science?

Can following the GL/CE affect how you carry out analyses (in either ASA or ACM terms) in your typical work? How/How not?

Is the SHA in Table 2.4.1 enough to ensure ethical analysis under either the ASA or the ACM uses of this term? Why/why not?

Discuss how the ASA and ACM are, or are not, fully aligned with "the scientist as a responsible member of society, contemporary ethical issues in (biomedical) research, and the environmental and societal impacts of scientific research". Consider "scientist" to mean anyone who utilizes the scientific method in the usual course of their work, even if they are employed in business or other (non-"science") domains.

Discuss how the ASA and ACM are, or are not, fully aligned with current policies (for your institution, organization, or lab/work team) on "conflict of interest".

Chapter 2.5
Interpretation

NOTE: Like "analysis", "interpretation" has different meanings for straight computation (ACM) and straight statistical (ASA) users. The meaning for statisticians and data scientists in terms of the ASA GLs are clearly referring to interpreting the analysis of data or in the case of meta-analysis, the interpretation of a collection of evidence (and data). The meaning of "interpretation" in the statistical sense will involve competent assessment of data, methods, and results. By contrast, because the ACM CE uses the term "analysis" (or related ideas) to refer to the analysis of risks, the evaluation (and implied analysis) of systems or their plans/designs, to the analysis of activities, or to the effects of computing on stakeholders, we consider that the coverage of the ACM CE on "analysis" sufficiently addresses the ACM's perspective on how the computing professional should utilize (interpret) the results of those analyses. This chapter, then, focuses only on the ASA GLs and their specific perspectives on interpretation, relating more particularly to the interpretation of statistical analyses. As the interpretation of statistical analysis is relevant in any computing system, the ASA GLs should be sufficiently informative for the ACM member/individual who utilizes the ACM CE to guide their decision making.

ASA Ethical Guidelines (GLs) for Statistical Practice: on interpretation

Overall: The first paragraph of the preamble to the GLs states that:

In these Guidelines, "statistical practice" includes activities such as: designing the collection of, summarizing, processing, analyzing, *interpreting*, or presenting, data; as well as model or algorithm development and deployment. Throughout these Guidelines, *the term "statistical practitioner" includes all those who engage in statistical practice, regardless of job title, profession, level, or field of degree."* (Emphasis added).

The Ethical Guidelines for Statistical Practice are intended to support ethical decision making, and one of the principal ways this happens is when appropriate methodology is applied to ethically sourced data. Then, the interpretations that are supported by the combination of methods, analysis, and results must also be ethically made. In the statement of Purpose of the

Guidelines, interpretation – even if this is not done by the individual who did the analyses – falls into the definition of *statistical practice*; whichever team member does the interpreting incurs the responsibility to do so ethically. Interpretations must be valid for decisions based upon them to actually support an informed society. **Invalid** interpretations of otherwise ethically executed analyses on ethically sourced data will *undermine*, not support, the informedness of society.

Moreover, the GLs are intended to be used by "all those who engage in statistical practice, regardless of job title, profession, level, or field of degree." Those who take the responsibility for interpreting the application of statistical methods bear a particular burden of ensuring that statistical results are interpreted ethically and in a manner consistent with methods and data. Purposefully misunderstanding/misinterpreting statistics is unethical and detrimental; purposefully misrepresenting results in pursuit of unethical ends is naturally inherently unethical. The ethical statistical practitioner promotes correct, supportable, interpretations and does not condone, tolerate, or promote the scientific misconduct that arises from misrepresentation of their analyses or results.

Principle A: Professional integrity and accountability

Professional integrity and accountability require taking responsibility for one's work. Ethical statistical practice supports valid and prudent decision making with appropriate methodology. The ethical statistical practitioner represents their capabilities and activities honestly, and treats others with respect.

The ethical statistical practitioner:

A2 Uses methodology and data that are valid, relevant, and appropriate, without favoritism or prejudice, and in a manner intended to produce valid, interpretable, and reproducible results.

A3 Does not knowingly conduct statistical practices that exploit vulnerable populations or create or perpetuate unfair outcomes.

A4 Opposes efforts to predetermine or influence the results of statistical practices, and resists pressure to selectively interpret data.

A9 Takes appropriate action when aware of deviations from these Guidelines by others.

Principle A, as well as these four identified elements, relate specifically to how important it is for the ethical practitioner to interpret the results of their (and others') work competently and transparently. Particularly in cases where investigators or data providers are inclined to predetermine or influence the analyses or results, the ethical statistical practitioner is vigilant about ensuring that any interpretations are specifically supported by the methods and results. Favoritism and prejudice in the interpretation of results may arise from, or create, conflicts of interest. Reasonable interpretations that do not over-extrapolate or over-generalize should be provided with every analysis report. The International Committee of Medical Journal Editors (http://www.icmje.org/recommendations/browse/roles-and-responsibilities/defining-the-role-of-authors-and-contributors.html), requires that any individual listed as an author on work must "... be accountable for all aspects of the work in ensuring that questions related to the accuracy or integrity of any part of the work are appropriately investigated and resolved". This external (to the GLs) definition of authorship underscores the role of the ethical statistical practitioner in ensuring that no collaborators, or employers, alter the interpretations beyond what the methods, data, and results will support. As previously stated, "Invalid interpretations of otherwise ethically executed analyses on ethically sourced data will *undermine*, not support, informedness".

The ethical statistical practitioner is accountable for the analyses and results, and ensures that their work produces valid, interpretable, and reproducible results. Invalid interpretations tend to be irreproducible, which wastes time and resources (as well as undermining, rather than supporting, informedness). Professional integrity as well as professional accountability (whether as an author or otherwise) require that interpretations are specifically supportable by the data and the methods. Ensuring that interpretations are valid and specific to the data, methods, and results can be perceived as falling outside the statistician's purview, but the statistician is professionally accountable for their work, and invalid, irreproducible interpretations of their work undermine professionalism and the profession (Principle F).

Principle B. Integrity of data and methods

The ethical statistical practitioner seeks to understand and mitigate known or suspected limitations, defects, or biases in the data or methods and communicates potential impacts on the interpretation, conclusions, recommendations, decisions, or other results of statistical practices.

The ethical statistical practitioner:

B1 Communicates data sources and fitness for use, including data generation and collection processes and known biases. Discloses and manages any conflicts of interest relating to the data sources.

Communicates data processing and transformation procedures, including missing data handling.

B2   Is transparent about assumptions made in the execution and interpretation of statistical practices including methods used, limitations, possible sources of error, and algorithmic biases. Conveys results or applications of statistical practices in ways that are honest and meaningful.

B3   Communicates the stated purpose and the intended use of statistical practices. Is transparent regarding a priori versus post hoc objectives and planned versus unplanned statistical practices. Discloses when multiple comparisons are conducted, and any relevant adjustments.

B4   Meets obligations to share the data used in the statistical practices, for example, for peer review and replication, as allowable. Respects expectations of data contributors when using or sharing data. Exercises due caution to protect proprietary and confidential data, including all data that might inappropriately harm data subjects.

B5   Strives to promptly correct substantive errors discovered after publication or implementation. As appropriate, disseminates the correction publicly and/or to others relying on the results.

B6   For models and algorithms designed to inform or implement decisions repeatedly, develops and/or implements plans to validate assumptions and assess performance over time, as needed. Considers criteria and mitigation plans for model or algorithm failure and retirement.

B7   Explores and describes the effect of variation in human characteristics and groups on statistical practice when feasible and relevant.

Every element in Principle B, as well as the Principle itself, relates directly or indirectly to interpretation. Full and transparent reporting of analyses and results is required to enable an unbiased reader to make supported interpretations and avoid drawing conclusions that are not supported/not valid. The integrity of the data and appropriateness of the analytic methods are undermined when invalid interpretations are permitted (if made by others) or made by the statistician. Misinterpretation and misrepresentation are perhaps the most common, and most detrimental, "substantive error" a practitioner will encounter.

Principle C. Responsibilities to stakeholders

Those who fund, contribute to, use, or are affected by statistical practices are considered stakeholders. The ethical statistical practitioner respects the interests of stakeholders while practicing in compliance with these Guidelines.

The ethical statistical practitioner:

C2 Regardless of personal or institutional interests or external pressures, does not use statistical practices to mislead any stakeholder.

C3 Uses practices appropriate to exploratory and confirmatory phases of a project, differentiating findings from each so the stakeholders can understand and apply the results.

C8 Prioritizes both scientific integrity and the principles outlined in these Guidelines when interests are in conflict.

The ethical statistical practitioner neither tolerates, condones, nor engages in statistical, professional, or scientific misconduct. In some contexts, the report of the statistician is handed over to another team member who will then summarize and draw conclusions based on that work. Misrepresentation or misinterpretation of results has the potential for far-reaching and negative harm to the scientific community and to the public, particularly in terms of undermining the public trust in the scientific enterprise. The ethical statistical practitioner's obligation to ensure that interpretations are directly supported by the data, methods, and results takes priority over responsibilities to provide the client or funder their desired results. As noted, informedness is weakened when results and analyses are misinterpreted; additionally, misinterpretation or misrepresentation of results represent scientific misconduct and are not professional. However, as the ethical statistical practitioner seeks to ensure that interpretations are appropriate, others (e.g., those who prioritize the desires of the funder or client over those of the scientific community and the public trust) may seek to intimidate or otherwise pressure the ethical statistical practitioner. Maintaining a professional perspective in the face of team pressures can be very challenging, and these GLs are intended to support the practitioner who seeks to prioritize stakeholder interests as they promote appropriate interpretation. When misinterpretation or misrepresentation of statistical practices are identified, scientific integrity (and the GLs) should be prioritized, so that misrepresentations are countered or challenged (e.g., following B5).

Principle D. Responsibilities to research subjects, data subjects, or those directly affected by statistical practices

The ethical statistical practitioner does not misuse or condone the misuse of data. They protect and respect the rights and interests of human and animal subjects. These responsibilities extend to those who will be directly affected by statistical practices.

The ethical statistical practitioner:

D6   Considers the impact of statistical practice on society, groups, and individuals. Recognizes that statistical practice could adversely affect groups or the public perception of groups, including marginalized groups. Considers approaches to minimize negative impacts in applications or in framing results in reporting.

D11   Does not conduct statistical practice that could reasonably be interpreted by subjects as sanctioning a violation of their rights. Seeks to use statistical practices to promote the just and impartial treatment of all individuals.

Ensuring valid and supportable/appropriate interpretations is the crux of the obligation of the ethical statistical practitioner to the research subjects or data contributors. Note that *purposeful misinterpretation or misrepresentation of statistical practices are a misuse of data.* This is one reason why the NASEM (2017) report identifies statistical misconduct as explicitly damaging and detrimental. The ethical statistical practitioner avoids the use of data that may violate data contributors' rights, or appear to sanction the violation of their rights. The ethical practitioner demonstrates their respect for the animals and humans whose data are analyzed by ensuring that interpretations of any and all analyses are valid and reproducible. Misinterpretations and misrepresentations that lead to irreproducible results effectively waste the data that were contributed, and constitute a failure in this responsibility to respect their rights and interests.

Principle E. Responsibilities to members of multidisciplinary teams

Statistical practice is often conducted in teams made up of professionals with different professional standards. The statistical practitioner must know how to work ethically in this

The ethical statistical practitioner:

E2   Prioritizes these Guidelines for the conduct of statistical practice in cases where ethical guidelines conflict.

E3   Ensures that all communications regarding statistical practices are consistent with these Guidelines. Promotes transparency in all statistical practices.

E4   Avoids compromising validity for expediency. Regardless of pressure on or within the team, does not use inappropriate statistical practicesE4

Avoids compromising validity for expediency. Regardless of pressure on or within the team, does not use inappropriate statistical practices.

The ethical statistical practitioner recognizes that other team members may not be familiar with, or feel obliged to follow, these GLs (even if they are also statistical practitioners), but **that does not affect the analyst's obligations to do so.** The ethical statistical practitioner neither tolerates, condones, nor engages in professional or scientific misconduct, or intimidation. These behaviors may be brought to bear when the ethical statistical practitioner tries to make sure that results are not misinterpreted or misrepresented. Maintaining a professional perspective in the face of team pressures can be very challenging, but ensuring that interpretations are valid and supported is an essential obligation of the ethical statistical practitioner, because "(s)ociety benefits from informed judgments supported by ethical statistical practice" (Purpose of the Guidelines). When results are misinterpreted or misrepresented, society is less informed, and judgments are not supported by statistical methods (because the results of those methods are misrepresented). This is true for the wider society (negatively impacting the public trust in the profession), the scientific community (negatively impacting the public trust in the scientific community), and for the profession.

Principle F. Responsibilities to fellow statistical practitioners and the Profession

Statistical practices occur in a wide range of contexts. Irrespective of job title and training, those who practice statistics have a responsibility to treat statistical practitioners, and the profession, with respect. Responsibilities to other practitioners and the profession include honest communication and engagement that can strengthen the work of others and the profession.

The ethical statistical practitioner:

F4 Promotes reproducibility and replication, whether results are "significant" or not, by sharing data, methods, and documentation to the extent possible.

F5 Serves as an ambassador for statistical practice by promoting thoughtful choices about data acquisition, analytic procedures, and data structures among non-practitioners and students. Instills appreciation for the concepts and methods of statistical practice.

The ethical statistical practitioner is respectful of other statisticians, focusing on the science, methodologies, and the reproducibility of interpretable results. As noted, per H2, the ethical statistical practitioner neither tolerates, condones, nor engages in statistical, professional, or scientific misconduct. Ethical interpretation promotes reproducible and rigorous results.

When interpretations of analyses are in conflict, thoughtful synthesis of diverse results will strengthen the discipline of statistics as well as supporting the interpretability and reproducibility of results. Permitting misrepresentation (misinterpretation – often *over*interpretation) of results not only undermines the informedness of society, it also indirectly weakens the discipline and may falsely lead non-statistical team members to expect to be able to draw any conclusions they want, contradicting most of the GL Principles (as outlined in Principle E and the Appendix).

Principle G. Responsibilities of leaders, supervisors, and mentors in statistical practice

Those leading, supervising, or mentoring statistical practitioners are expected to:

G1 Ensure appropriate statistical practice that is consistent with these Guidelines. Protect the statistical practitioners who comply with these Guidelines, and advocate for a culture that supports ethical statistical practice.

G5 Establish a culture that values validation of assumptions, and assessment of model/algorithm performance over time and across relevant subgroups, as needed. Communicate with relevant stakeholders regarding model or algorithm maintenance, failure, or actual or proposed modifications.

As noted, ensuring that correct and appropriate interpretations are made based on the ethical statistical practitioner's work can sometimes create a uniquely difficult burden in following the ASA GLs. Statistics practitioners in leadership roles should support each analyst's efforts to ensure that interpretations are appropriate, and ensure that there are no undue pressures placed on the statistical practitioner to comply with misrepresentation of results. Complete and transparent documentation of planning and design by the ethical statistical practitioner will create a record of the work that was actually done and the interpretations that were actually supported by that work, and in the event that someone else's misrepresentation of that work comes to light, at least the original documentation will correctly reflect the statistician's intent. The ethical statistical practitioner as a team leader ensures that only appropriate interpretations are made based on all statistical work, promoting G1 and G5, and protecting the professional freedom and responsibility of those who comply with these GLs.

Principle H. Responsibilities regarding potential misconduct

Th ethical statistical practitioner:

> H2 Avoids condoning or appearing to condone statistical, scientific, or professional misconduct. Encourages other practitioners to avoid misconduct or the appearance of misconduct.

Misrepresentation of results can constitute fabrication, fraud, or a false claim that might be actionable according to the US Department of Justice. Research misconduct is specifically defined to include deliberate misrepresentation of an analysis (any research). Differences of opinion on appropriate conclusions and inferences, or honest errors in interpretation, do not constitute misconduct. In a workplace where ethical statistical practice is supported (see Principle G), transparency and completeness in reporting will also strengthen the perception of valid interpretations. Whenever pressure, intimidation, or other unprofessional actions are taken to divert or suppress complaints about misrepresentation or misinterpretation of results, it increases the likelihood of a deliberate misrepresentation. While "research misconduct" and "false claims" are actionable only when these are done in bids for federal monies, deliberate misrepresentation and misinterpretation of statistical work is professional and scientific misconduct and should be reported, avoided, and not tolerated.

Although we do not include a formal assessment of the ACM CE in terms of interpretation, it is worthwhile to note that the computing professional may interpret their analysis of behavior or systems in a way that results in the identification of violations of ACM Principles. While the interpretation is not discussed specifically in the ACM CE, the actions taken to address the interpretation require careful consideration so that the appropriate action can be taken:

> 1.2 A computing professional has an additional obligation to report any signs of system risks that might result in harm. If leaders do not act to curtail or mitigate such risks, it may be necessary to "blow the whistle" to reduce potential harm. However, capricious or misguided reporting of risks can itself be harmful. Before reporting risks, a computing professional should carefully assess relevant aspects of the situation.

2.3 Rules that are judged unethical should be challenged. A rule may be unethical when it has an inadequate moral basis or causes recognizable harm. A computing professional should consider challenging the rule through existing channels before violating the rule. A computing professional who decides to violate a rule because it is unethical, or for any other reason, must consider potential consequences and accept responsibility for that action.

2.5 A system for which future risks cannot be reliably predicted requires frequent reassessment of risk as the system evolves in use, or it should not be deployed. Any issues that might result in major risk must be reported to appropriate parties.

2.8 Under exceptional circumstances a computing professional may use unauthorized access to disrupt or inhibit the functioning of malicious systems; extraordinary precautions must be taken in these instances to avoid harm to others.

2.9 In cases where misuse or harm are predictable or unavoidable, the best option may be to not implement the system.

3.4 Designing or implementing processes that deliberately or negligently violate, or tend to enable the violation of, the Code's principles is ethically unacceptable.

3.7 When appropriate standards of care do not exist, computing professionals have a duty to ensure they are developed.

4.1 Computing professionals who recognize breaches of the Code should take actions to resolve the ethical issues they recognize, including, when reasonable, expressing their concern to the person or persons thought to be violating the Code.

4.2 ACM members who recognize a breach of the Code should consider reporting the violation to the ACM.

The stakeholder analysis (SHA) table below shows some harms, benefits, unknown outcomes, and unknowable outcomes, relating to following the GLs and CE in interpretation.

Potential result: Stakeholder:	HARM	BENEFIT	UNKNOWN	UNKNOWABLE
YOU	Takes time, adds effort to ensure that valid interpretations (by analyst) are retained and accompany all analyses.	Good statistical practice requires valid interpretations. Compliance strengthens your reputation for excellence.	Invalid interpretations may undermine study design – irreproducible results will not yield robust answers to scientific questions.	Harms, rather than intended benefits, of studies with invalid interpretations may accrue and future work may be misdirected or unfunded.
Your boss/client	Valid interpretations may not be what was desired/expected. Invalid interpretations are unlikely to reproduce and are not rigorous, so client will not be able to capitalize on them.	Valid interpretations allow client to stop investing time/effort in negative results, and will strengthen reliability and confidence in positive ones.	Invalid interpretations may bias estimates and effect sizes, derailing future research or analyses.	
Unknown individuals	Invalid interpretations can lead to bias (and incorrect reports) in the scientific literature.		Harms, rather than intended benefits, of studies with invalid interpretations may accrue and future work may be misdirected or unfunded.	

Potential result: Stakeholder:	HARM	BENEFIT	UNKNOWN	UNKNOWABLE
Employer	Valid interpretations may not be what was desired/ expected. Invalid interpretations are unlikely to reproduce and are not rigorous, so employer will not be able to capitalize on them.	Valid interpretations allow employer to avoid investing (further) in negative results, and will strengthen reliability and confidence in positive ones.		Could limit business/profit in the short run but transparency and validity of interpretations should strengthen the employer's informedness and future judgments.
Colleagues	May require challenges to colleagues that do not support valid interpretations (in favour of desired ones)	Valid interpretations are reproducible and rigorous, promoting trust in statistical analysis and results.		
Profession	Invalid interpretations undermine the profession and the credibility of all practitioners.	Transparency supports trust, valid interpretations support informedness, scientific excellence, and good statistical practice.	Modeling the core features of "good statistical practice" may positively influence future practitioners.	

Potential result:	HARM	BENEFIT	UNKNOWN	UNKNOWABLE
Stakeholder:				
Public/public trust	Invalid interpretations, even if the analyses are ethically/competently done, undermine the informedness of society and any decisions the statistical analyses were intended to support.	Valid interpretations strengthen transparency and the public trust in statistical practice.	Transparency supports trust, valid interpretations support public informedness, scientific excellence, and good statistical practice.	Invalid interpretations may cause harms immediately and continuously if uncorrected or perpetuated. They may also confuse the public (particularly when describing risks or harms, even incorrectly).

Table 2.5.1. *Stakeholder Analysis template: Interpretation (ASA GLs ONLY)*

Questions for Discussion:

What are the most striking features of the SHA for the task of interpretation (given the ASA GLs)? Are there particular results in the SHA table that stand out because they are A) highly predictable? B) unexpected? C) have possibly unexpectedly different (or similar) benefits or harms to different stakeholders?

Can following the ASA GLs affect how you interpret analyses in your typical work? How/How not? Do you think the ASA GLs can support interpretation by a computing professional who is working in data science?

Is the SHA in Table 2.5.1 enough to ensure ethical interpretation? Why/why not?

Discuss how the ASA GLs (and to the extent that it is applicable, ACM CE) are, or are not, fully aligned with "responsible authorship and publication". How is this relevant for the task of "interpretation"?

Discuss how the ASA GLs (and ACM CE to the extent that it is relevant) are, or are not, fully aligned with current policies (for your institution, organization, or lab/work team) on "mentor/mentee responsibilities and relationships". How is this relevant for the task of "interpretation"?

Discuss how the ASA GLs input on interpretation are, or are not, fully aligned with current policies (for your institution, organization, or lab/work team) relating to "research misconduct and policies for handling misconduct". How is this relevant for the task of "interpretation"?

Chapter 2.6
Documenting your work

ASA Ethical Guidelines (GLs) for Statistical Practice: on documenting your work

Overall: The preamble to the GLs states that:

"In these Guidelines, "statistical practice" includes activities such as: designing the collection of, *summarizing*, processing, analyzing, interpreting, or *presenting*, data; as well as model or algorithm development and deployment. Throughout these Guidelines, the term "statistical practitioner" includes *all those who engage in statistical practice, regardless of job title, profession, level, or field of degree.*" (Emphasis added.)

The ethical practitioner documents their work so that it, and by extension, the work of those around them, will be transparent, reproducible, and valid. While not all practitioners will be able to share the details of their work, due to work-related confidentiality considerations for example, the ethical practitioner is aware that failures to document their work fully and correctly can have long-term, and unpredictable, negative effects on their future work as well as those who may depend on their work in other areas. For those who can share their work with other practitioners and with the wider scientific community, the ethical practitioner acts in good faith to promote ethical work by others, by documenting their own work carefully and completely.

Principle A: Professional integrity and accountability

The ethical statistical practitioner:

- A2 Uses methodology and data that are valid, relevant, and appropriate, without favoritism or prejudice, and in a manner intended to produce valid, interpretable, and reproducible results.
- A3 Does not knowingly conduct statistical practices that exploit vulnerable populations or create or perpetuate unfair outcomes.
- A4 Opposes efforts to predetermine or influence the results of statistical practices, and resists pressure to selectively interpret data.

A5 Accepts full responsibility for their own work; does not take credit for the work of others; and gives credit to those who contribute. Respects and acknowledges the intellectual property of others.

Principle A has four elements that most directly relate to the importance of competently and transparently documenting all work. Documentation of assumptions and methods support the intention to produce valid, interpretable, and reproducible results (A2), and also give the practitioner – and peer reviewers – the opportunity to ensure that no exploitation or unfairness is created or perpetuated by the work (A3). Accurate documentation also represents efforts to resist undue influences (A4), and all of these are essential features of how the ethical practitioner "accepts full responsibility for their own work." (A5) You might consider whether documentation is sufficient for a peer to review it and consider it complete – even if it is not possible to share with peers or submit for peer review – in order to determine whether the work is sufficiently documented. Considering whether another practitioner would consider the documentation sufficient can be informative to the individual who seeks to ensure their documentation is sufficiently informative and transparent.

Principle B. Integrity of data and methods

The ethical statistical practitioner seeks to understand and mitigate known or suspected limitations, defects, or biases in the data or methods and communicates potential impacts on the interpretation, conclusions, recommendations, decisions, or other results of statistical practices.

The ethical statistical practitioner:

B1 Communicates data sources and fitness for use, including data generation and collection processes and known biases. Discloses and manages any conflicts of interest relating to the data sources. Communicates data processing and transformation procedures, including missing data handling.

B2 Is transparent about assumptions made in the execution and interpretation of statistical practices including methods used, limitations, possible sources of error, and algorithmic biases. Conveys results or applications of statistical practices in ways that are honest and meaningful.

B3 Communicates the stated purpose and the intended use of statistical practices. Is transparent regarding a priori versus post hoc objectives and planned versus unplanned statistical practices. Discloses when multiple comparisons are conducted, and any relevant adjustments.

B4 Meets obligations to share the data used in the statistical practices, for example, for peer review and replication, as allowable. Respects expectations of data contributors when using or sharing data. Exercises due caution to protect proprietary and confidential data, including all data that might inappropriately harm data subjects.

B5 Strives to promptly correct substantive errors discovered after publication or implementation. As appropriate, disseminates the correction publicly and/or to others relying on the results.

B6 For models and algorithms designed to inform or implement decisions repeatedly, develops and/or implements plans to validate assumptions and assess performance over time, as needed. Considers criteria and mitigation plans for model or algorithm failure and retirement.

B7 Explores and describes the effect of variation in human characteristics and groups on statistical practice when feasible and relevant.

Every element in Principle B relates specifically and/or indirectly to the full and transparent documentation of statistical work. Correct and complete documentation, even if it cannot be shared, promotes transparency and all of the ethical obligations around communication with respect to the data and methods utilized.

Principle C. Responsibilities to stakeholders

Those who fund, contribute to, use, or are affected by statistical practices are considered stakeholders. The ethical statistical practitioner respects the interests of stakeholders while practicing in compliance with these Guidelines.

The ethical statistical practitioner:

C1 Seeks to establish what stakeholders hope to obtain from any specific project. Strives to obtain sufficient subject-matter knowledge to conduct meaningful and relevant statistical practice.

C2 Regardless of personal or institutional interests or external pressures, does not use statistical practices to mislead any stakeholder.

C3 Uses practices appropriate to exploratory and confirmatory phases of a project, differentiating findings from each so the stakeholders can understand and apply the results.

C4 Informs stakeholders of the potential limitations on use and re-use of statistical practices in different contexts and offers guidance and alternatives, where appropriate, about scope, cost, and precision considerations that affect the utility of the statistical practice.

C5 Explains any expected adverse consequences from failing to follow through on an agreed-upon sampling or analytic plan.

C6 Strives to make new methodological knowledge widely available to provide benefits to society at large. Presents relevant findings, when possible, to advance public knowledge.

C7 Understands and conforms to confidentiality requirements for data collection, release, and dissemination and any restrictions on its use established by the data provider (to the extent legally required). Protects the use and disclosure of data accordingly. Safeguards privileged information of the employer, client, or funder.

C8 Prioritizes both scientific integrity and the principles outlined in these Guidelines when interests are in conflict.

Because stakeholder perspectives can vary – and even be opposed – for the same project, documentation may be one of the more complicated tasks to apply KSAs 1 and 2 to. The documentation may be, include, or even be limited to, internal reports. Even when documentation of statistical practices cannot be shared publicly, internal documentation is still vitally important because it allows a reader to recognize that objectives (C1) are consistent with practices, such that no stakeholder can be misled (C2). In some cases, because it will not be made public, internal documentation can be particularly vulnerable to unethical reporting practices; the ethical practitioner guards against these to the extent possible. Even for internal use only, the ethical practitioner has responsibilities to fully and completely document their work – including possible options that were/could be taken, and deviations from agreed-upon plans. In documentation, whether or not it can be shared or peer-reviewed, it is essential to *not include* privileged information (C6-C7); conflicts of interest may arise –and must be managed – when the full documentation of work seems to require that some privileged information be included. Possibly the biggest challenge to ethical practice is balancing C2 ("Regardless of personal or institutional interests or external pressures, does not use statistical practices to mislead any stakeholder") and C6 ("Strives to make new methodological knowledge widely available to provide benefits to society at large. Presents relevant findings, when possible, to advance public knowledge."). C8 offers scientific integrity and following the Guideline Principles as the priorities whenever stakeholder interests are in conflict, meaning that C2 (do not mislead any stakeholder) must be prioritized in terms of questions about documentation. Just as the practitioner might consider whether their documentation (even if it cannot be shared) is sufficient for another practitioner to replicate their work, considering whether the documentation is sufficient to ensure that no stakeholder would be misled by results can also be used as a test for the sufficiency of documentation.

Principle D. Responsibilities to research subjects, data subjects, or those directly affected by statistical practices

The ethical statistical practitioner:

D6 Considers the impact of statistical practice on society, groups, and individuals. Recognizes that statistical practice could adversely affect groups or the public perception of groups, including marginalized groups. Considers approaches to minimize negative impacts in applications or in framing results in reporting.

D7 Refrains from collecting or using more data than is necessary. Uses confidential information only when permitted and only to the extent necessary. Seeks to minimize the risk of re-identification when sharing de-identified data or results where there is an expectation of confidentiality. Explains any impact of de-identification on accuracy of results.

D8 To maximize contributions of data subjects, considers how best to use available data sources for exploration, training, testing, validation, or replication as needed for the application. The ethical statistical practitioner appropriately discloses how the data is used for these purposes and any limitations.

Three elements of Principle D relate specifically to documenting your work: whenever communication of methods, assumptions, or results is a function of the task, any GL Principle or element discussing reporting, communication, or documentation is also relevant. Elements D6, D7, and D8 discuss how documentation helps the ethical practitioner to accomplish their responsibilities to those who contribute data, or those who are directly affected by statistical practices.

Principle E. Responsibilities to members of multidisciplinary teams

The ethical statistical practitioner:

E3 Ensures that all communications regarding statistical practices are consistent with these Guidelines. Promotes transparency in all statistical practices.

E4 Avoids compromising validity for expediency. Regardless of pressure on or within the team, does not use inappropriate statistical practices.

In addition to transparent and complete documentation of what was done, the ethical practitioner also ensures that what they will be reporting was ethically designed and executed. The ethical obligations to document what data were used, where it was obtained, and what exactly was done (following Principle B) are embedded throughout practice, partly to discharge our Responsibilities to research subjects, data subjects, or those directly affected by statistical practices (D) and our Responsibilities to our team members (E). While the documentation must be complete, it is essential to keep in mind the applicable rules for the protection of those whose data were obtained (D1) and confidentiality obligations (C7). Transparency can only be achieved with complete and correct documentation (E3), and this should be prioritized irrespective of pressures within or on the team (E4).

Principle F. Responsibilities to fellow statistical practitioners and the Profession

Statistical practices occur in a wide range of contexts. Irrespective of job title and training, those who practice statistics have a responsibility to treat statistical practitioners, and the profession, with respect. Responsibilities to other practitioners and the profession include honest communication and engagement that can strengthen the work of others and the profession.

The ethical statistical practitioner:

F4 Promotes reproducibility and replication, whether results are "significant" or not, by sharing data, methods, and documentation to the extent possible.

F5 Serves as an ambassador for statistical practice by promoting thoughtful choices about data acquisition, analytic procedures, and data structures among non-practitioners and students. Instills appreciation for the concepts and methods of statistical practice.

Documentation of our work is embedded in our responsibilities to other practitioners, and to anyone who would seek to replicate our work. Ensuring documentation that is useable, and encouraging documentation to be as full as possible by all practitioners, are two key parts of ethical practice (F4). Part of "promoting thoughtful choices about" statistical practice is the documentation of how this is demonstrated by the ethical practitioner (F5).

Principle G. Responsibilities of leaders, supervisors, and mentors in statistical practice

Those leading, supervising, or mentoring statistical practitioners are expected to:

G1 Ensure appropriate statistical practice that is consistent with these Guidelines. Protect the statistical practitioners who comply with these Guidelines, and advocate for a culture that supports ethical statistical practice.

G5 Establish a culture that values validation of assumptions, and assessment of model/algorithm performance over time and across relevant subgroups, as needed. Communicate with relevant stakeholders regarding model or algorithm maintenance, failure, or actual or proposed modifications.

The ethical practitioner follows the ASA GLs and documents their work accordingly, so the ethical leader in statistical practice promotes documentation as well (G1/G5). Whenever the workplace suppresses this documentation, or if unethical reporting is encouraged (e.g., insufficiently detailed for others to replicate the work, or incorrect so that replications will fail), the ethical practitioner recognizes that these behaviors are contrary to the GLs, and should not be allowed/followed. The ethical practitioner documents their work and supports the profession as well as the freedom and responsibility of statistical practitioners who comply with these GLs (G1).

Principle H. Responsibilities regarding potential misconduct

The ethical statistical practitioner:

H2 Avoids condoning or appearing to condone statistical, scientific, or professional misconduct. Encourages other practitioners to avoid misconduct or the appearance of misconduct.

The ethical practitioner engages in, and directly and indirectly supports, full and correct documentation of all work. Ensuring the documentation is complete and correct, and is limited in terms of proprietary or confidential information that should not be shared (see discussion of Principle C) helps the practitioner to achieve their obligations towards ethical conduct (statistical, scientific, and professional). Documentation that is complete and correct also avoids both condoning and engaging in incompetent and unprofessional behavior. Intentionally misleading a reader or future user of documentation (or the method/technique that is documented) undermines the statistical profession as well as demonstrating a lack of professional integrity, and is unprofessional.

ACM Code of Ethics: on documenting your work

Overall: The Preamble states,

The entire computing profession benefits when the ethical decision-making process is accountable to and transparent to all stakeholders. Open discussions about ethical issues promote this accountability and transparency.

In order to be transparent, and to enable accountability and open discussions about decisions that are made, documentation of both one's work and the decisions that were involved should be encouraged.

1. General Ethical Principles

1.1 Computing professionals should consider whether the results of their efforts will respect diversity, will be used in socially responsible ways, will meet social needs, and will be broadly accessible.

1.2 A computing professional has an additional obligation to report any signs of system risks that might result in harm. If leaders do not act to curtail or mitigate such risks, it may be necessary to "blow the whistle" to reduce potential harm. However, capricious or misguided reporting of risks can itself be harmful. Before reporting risks, a computing professional should carefully assess relevant aspects of the situation.

1.3 A computing professional should be transparent and provide full disclosure of all pertinent system capabilities, limitations, and potential problems to the appropriate parties.
Computing professionals should be forthright about any circumstances that might lead to either real or perceived conflicts of interest or otherwise tend to undermine the independence of their judgment.

1.4 Failure to design for inclusiveness and accessibility may constitute unfair discrimination.

1.5 Computing professionals should therefore credit the creators of ideas, inventions, work, and artifacts, and respect copyrights, patents, trade secrets, license agreements, and other methods of protecting authors' works.

1.6 Computing professionals should establish transparent policies and procedures that allow individuals to understand what data is being collected and how it is being used, to give informed consent for automatic data collection, and to review, obtain, correct inaccuracies in, and delete their personal data.

Only the minimum amount of personal information necessary should be collected in a system. The retention and disposal periods for that information should be clearly defined, enforced, and communicated to data subjects. Personal information gathered for a specific purpose should not be used for other purposes without the person's consent. Merged data collections can compromise privacy features present in the original collections. Therefore, computing professionals should take special care for privacy when merging data collections.

1.7 Computing professionals should protect confidentiality except in cases where it is evidence of the violation of law, of organizational regulations, or of the Code. In these cases, the nature or contents of that information should not be disclosed except to appropriate authorities. A computing professional should consider thoughtfully whether such disclosures are consistent with the Code.

Many of the ACM CE General Ethical Principles (1) relate to transparency/accountability practices, or other allusions to communication that ultimately require documentation of all work. Failures to document one's computing work may result in problems when following ACM 1.5 and claiming one created something in case another person, professional, or author claims it. These Principles do not explicitly mention the documentation of work but there are distinctive suggestions that, for example, the computing professional that follows ACM 1.7 and discloses some information might want/need to document how they "consider(-ed) thoughtfully whether such disclosures are consistent with the Code". The establishment of a policy or other program/system should be documented – particularly in terms of how designs are intended to avoid bias and promote accessibility. The documentation requirement for computing professionals might be slightly different than it is for the statistical professional if re-creation of the computing system is not essential (as it often is for statistics). However, the ACM General

Ethical Principle (1) does offer guidance on what *should be* documented by computing professionals, in order to ensure transparency and accountability.

2. Professional Responsibilities

2.1 Computing professionals should respect the right of those involved to transparent communication about the project.

2.2 Professional competence starts with technical knowledge and with awareness of the social context in which their work may be deployed. Professional competence also requires skill in communication, in reflective analysis, and in recognizing and navigating ethical challenges.

2.3 Rules that are judged unethical should be challenged. A rule may be unethical when it has an inadequate moral basis or causes recognizable harm.

A computing professional who decides to violate a rule because it is unethical, or for any other reason, must consider potential consequences and accept responsibility for that action.

2.5 Computing professionals are in a position of trust, and therefore have a special responsibility to provide objective, credible evaluations and testimony to employers, employees, clients, users, and the public. Computing professionals should strive to be perceptive, thorough, and objective when evaluating, recommending, and presenting system descriptions and alternatives. Extraordinary care should be taken to identify and mitigate potential risks in machine learning systems. A system for which future risks cannot be reliably predicted requires frequent reassessment of risk as the system evolves in use, or it should not be deployed. Any issues that might result in major risk must be reported to appropriate parties.

2.6 A computing professional is responsible for evaluating potential work assignments. This includes evaluating the work's feasibility and advisability, and making a judgment about whether the work assignment is within the professional's areas of competence. If at any time before or during the work assignment the professional identifies a lack of a necessary expertise, they must disclose this to the employer or client.

2.8 Under exceptional circumstances a computing professional may use unauthorized access to disrupt or inhibit the functioning of malicious systems; extraordinary precautions must be taken in these instances to avoid harm to others.

2.9 Computing professionals should perform due diligence to ensure the system functions as intended, and take appropriate action to secure resources against accidental and intentional misuse, modification, and denial of service. As threats can arise and change after a system is deployed, computing professionals should integrate mitigation techniques and policies, such as monitoring, patching, and vulnerability reporting.

In cases where misuse or harm are predictable or unavoidable, the best option may be to not implement the system.

The responsibilities of the ethical computing professional involve extensive contextual and stakeholder analysis, and also documentation (of your consideration) of contextual and competency issues and how these are/were resolved. In these senses (responding to ACM Principle 2, Professional Responsibilities), the documentation of how a system works/will or should work, and of the decisions going into the work, as drawn up by the ethical practitioner will represent the communication or analyses that support or precede decisions and actions. As the Preamble states, "The entire computing profession benefits when the ethical decision-making process is accountable to and transparent to all stakeholders", and this can only be accomplished with documentation of this process. Following rules or prioritizing following the ACM CE instead of other rules should be documented, and in this sense, the "documentation of the computing professional's work" relates to outlining how they arrived at specific features of the decisions that they made. Because each decision can have an impact on both stakeholders and the future of a project, it is important to document the "ethical analysis" – the thinking behind ethical decision-making, throughout the diverse aspects of computing practice. When the computing professional determines that misuse or harm are too likely, and that a system should not be designed or implemented, that decision and the reasoning behind it may be essential in the thinking or reasoning of others who might not have fully considered all of the risks for harm. Finally, decisions that automated systems make are increasingly scrutinized [22], necessitating full documentation of how the systems function (if the individual decisions cannot explicitly be determined). Thus, the fullest documentation of both system design/function and ethical analysis are essential in the "Open discussions about ethical issues (that) promote this accountability and transparency."

---

[22] General Principle 5 (p. 4), "transparency": The basis of a particular A/IS decision should always be discoverable" (IEEE 2019).

3. Professional Leadership Principles

3.1 People—including users, customers, colleagues, and others affected directly or indirectly— should always be the central concern in computing. The public good should always be an explicit consideration when evaluating tasks associated with research, requirements analysis, design, implementation, testing, validation, deployment, maintenance, retirement, and disposal. Computing professionals should keep this focus no matter which methodologies or techniques they use in their practice.

3.4 Leaders should pursue clearly defined organizational policies that are consistent with the Code and effectively communicate them to relevant stakeholders. In addition, leaders should encourage and reward compliance with those policies, and take appropriate action when policies are violated. Designing or implementing processes that deliberately or negligently violate, or tend to enable the violation of, the Code's principles is ethically unacceptable.

3.5 Computing professionals should be fully aware of the dangers of oversimplified approaches, the improbability of anticipating every possible operating condition, the inevitability of software errors, the interactions of systems and their contexts, and other issues related to the complexity of their profession—and thus be confident in taking on responsibilities for the work that they do.

Again, the accomplishment of ACM CE Principle 3 requires documentation of the thinking that underpins the policies and decisions that the leader in computing will make. Without documentation of the ethical analyses that support 3.1, 3.4 and 3.5 in particular, communication of the decisions will simply be assertions or statements, and may not have the intended positive and far-reaching effects that were intended. Engaging all computing professionals in the process of ethical analysis can help them to "grow as professionals" (ACM 3.5), adding particular importance to the documentation of this thinking for the leader in computing.

4. Compliance with the Code

4.1 Computing professionals who recognize breaches of the Code should take actions to resolve the ethical issues they recognize, including, when reasonable, expressing their concern to the person or persons thought to be violating the Code.

4.2 ACM members who recognize a breach of the Code should consider reporting the violation.

As discussed for other ACM Principles, the identification of failures to comply with the ACM CE will require careful thought – and documentation of all considerations in making a decision- about notifying either the individual or the ACM itself about suspected violations of the ACM CE.

The stakeholder analysis (SHA) table below shows some harms, benefits, unknown outcomes, and unknowable outcomes, relating to following the GLs and CE in documenting your work.

Potential result: Stakeholder:	HARM	BENEFIT	UNKNOWN	UNKNOWABLE
YOU	Takes time, adds effort (to fully document plans/evaluations/ systems); may facilitate use of work that is unauthorized/ does not acknowledge original creator	Simplifies evaluations of the work (ASA) or system, and updates (ACM); adds transparency and accountability.	ACM: New harms/risks may be detectable more quickly (and addressed) with full documentation. ASA: replications are facilitated.	
Your boss/client	Takes time	Creates transparency and accountability of statistician/ computing professional work		
Unknown individuals		Creates transparency and accountability		
Employer	Takes time	Demonstrates commitment to transparency & accountability.	ACM: New harms/risks may be detectable more quickly (and addressed) with full documentation. ASA: replications are facilitated.	ACM: Harms must be remediated, so could limit business/profit (although ASA/ACM would not consider this ethical justification to forego documentation).

Potential result: Stakeholder:	HARM	BENEFIT	UNKNOWN	UNKNOWABLE
Colleagues	May add time (in reviewing)	Supports transparency, accountability, and replicability. Promotes peer evaluation.		
Profession	Takes time, adds effort (to fully document plans/evaluations/ systems); may facilitate use of work that is unauthorized/ does not acknowledge original creator	Simplifies evaluations of the work (ASA) or system, and updates (ACM); adds transparency and accountability. Promotes and facilitates peer evaluation.	ACM: encourages careful review, so new harms/risks may be detectable more quickly (and addressed). ASA: transparency and validity of analyses and interpretations are facilitated.	
Public/public trust	Without full documentation of work/thought processes, public trust in systems or their results, and the profession (ACM/ASA) are compromised. Perpetuating black box mentality does not help, and explicitly does not inform, society.	Transparency and accountability – and the potential to understand how decisions are made based on automation and systems - supports public trust in the profession and its work.	Transparency supports trust, even if errors are discovered; their discoverability arises only and specifically from documentation, transparency, and accountability.	

Table 2.6.1. *Stakeholder Analysis template: Documenting your work*

Questions for Discussion:

Discuss similarities and differences between the ASA GLs and ACM CE principles that are discussed above, in terms of how they apply to documenting your work.

What are the most striking similarities (e.g., what is unexpectedly similar)? Are there particular differences that stand out because they are A) highly predictable? B) unexpected? C) have different benefits or harms to different stakeholders?

Rank GL/CE elements from most to least important to documenting your work. What does the rankability mean for you or for the discipline of statistics and data science? Is there one element or Principle (on either GL or CE) that is most challenging when it comes to documentation? How can this/these be addressed ethically?

Can following the GL/CE affect how you document your work in your typical project? How/How not?

Is the SHA in Table 2.6.1 enough to ensure ethical practice relating to documenting your work? Why/why not?

Discuss how the ASA and ACM are, or are not, fully aligned with current policies (for your institution, organization, or lab/work team) on "collaborative research, including collaborations with industry".

Discuss how the ASA and ACM are, or are not, supportive of ethical "peer review".

Chapter 2.7
Reporting your results/communication

ASA Ethical Guidelines (GLs) for Statistical Practice: <u>on</u> reporting and communication

Overall: The Purpose of the GLs states that:

"In these Guidelines, "statistical practice" includes activities such as: designing the collection of, *summarizing,* processing, analyzing, interpreting, or *presenting,* data; as well as model or algorithm development and deployment. Throughout these Guidelines, the term "statistical practitioner" includes *all those who engage in statistical practice, regardless of job title, profession, level, or field of degree."* (Emphasis added.)

The ethical practitioner documents their work so that it, and by extension, the work of those around them, will be transparent, reproducible, and valid. While not all practitioners will be able to share the details of their work, due to work-related confidentiality considerations for example, the ethical practitioner is aware that failures to communicate transparently and honestly can have long-term, and unpredictable, negative effects on their future work as well as those who may depend on their work in other areas. For those who can share their work with other practitioners and with the wider scientific community, the ethical practitioner acts in good faith to promote ethical work by others, by communicating transparently, and ensuring that those to whom their work is presented, communicated, or reported are not misled. An important aspect of the description of all those to whom the Guidelines apply is that oftentimes, the statistical practitioner does not formulate the final report and communication. Any individual who takes a technical report of statistical practices and summarizes and/or presents it to others is obliged to follow these Guidelines, *because their role with the work qualifies them as a "statistical practitioner".*

Principle A: Professional integrity and accountability

The ethical statistical practitioner:

> A2   Uses methodology and data that are valid, relevant, and appropriate, without favoritism or prejudice, and in a manner intended to produce valid, interpretable, and reproducible results.

A3 Does not knowingly conduct statistical practices that exploit vulnerable populations or create or perpetuate unfair outcomes.

A4 Opposes efforts to predetermine or influence the results of statistical practices, and resists pressure to selectively interpret data.

A5 Accepts full responsibility for their own work; does not take credit for the work of others; and gives credit to those who contribute. Respects and acknowledges the intellectual property of others.

The ASA asserts that the ethical practitioner strives to do and support reporting and communication that is transparent, reproducible, and valid. Moreover, the ethical practitioner both acts in a manner consistent with the GLs in their own reporting, and also encourages others to do so. These four elements of Principle A relate most specifically to how reporting and communicating must be done (i.e., competently and transparently). Note that it is inconsistent with the Guidelines to take an ethically generated output of statistical practice (e.g., an analysis) and report it in a way that is not transparent, may mislead readers, or is otherwise selective or inappropriate. Even giving full credit to the creator of the outputs, and then misrepresenting these outputs, is a violation of the intent and the letter of these Guidelines.

Principle B. Integrity of data and methods

The ethical statistical practitioner seeks to understand and mitigate known or suspected limitations, defects, or biases in the data or methods and communicates potential impacts on the interpretation, conclusions, recommendations, decisions, or other results of statistical practices.

The ethical statistical practitioner:

B1 Communicates data sources and fitness for use, including data generation and collection processes and known biases. Discloses and manages any conflicts of interest relating to the data sources. Communicates data processing and transformation procedures, including missing data handling.

B2 Is transparent about assumptions made in the execution and interpretation of statistical practices including methods used, limitations, possible sources of error, and algorithmic biases. Conveys results or applications of statistical practices in ways that are honest and meaningful.

B3 Communicates the stated purpose and the intended use of statistical practices. Is transparent regarding a priori versus post hoc objectives and planned versus unplanned statistical practices. Discloses when multiple comparisons are conducted, and any relevant adjustments.

B5 Strives to promptly correct substantive errors discovered after publication or implementation. As appropriate, disseminates the correction publicly and/or to others relying on the results.

B6 For models and algorithms designed to inform or implement decisions repeatedly, develops and/or implements plans to validate assumptions and assess performance over time, as needed. Considers criteria and mitigation plans for model or algorithm failure and retirement.

B7 Explores and describes the effect of variation in human characteristics and groups on statistical practice when feasible and relevant.

All but one (B4) element in Principle B relates specifically to ethical reporting and communication practices, in terms of doing the reporting and supporting ethical reporting/communication through transparent reporting and effective communication. Note that B6 involves communication within a team, to ensure that ongoing performance assessment is planned and implemented, while B5 relates more to communication with other statistical practitioners or end users of the outputs of statistical practice. B7 relates more to the scientific community, whereas B1-B3 are relevant in all contexts.

Principle C. Responsibilities to stakeholders

Those who fund, contribute to, use, or are affected by statistical practices are considered stakeholders. The ethical statistical practitioner respects the interests of stakeholders while practicing in compliance with these Guidelines.

The ethical statistical practitioner:

C1 Seeks to establish what stakeholders hope to obtain from any specific project. Strives to obtain sufficient subject-matter knowledge to conduct meaningful and relevant statistical practice.

C2 Regardless of personal or institutional interests or external pressures, does not use statistical practices to mislead any stakeholder.

C3 Uses practices appropriate to exploratory and confirmatory phases of a project, differentiating findings from each so the stakeholders can understand and apply the results.

C4 Informs stakeholders of the potential limitations on use and re-use of statistical practices in different contexts and offers guidance and alternatives, where appropriate, about scope, cost, and precision considerations that affect the utility of the statistical practice.

C5 Explains any expected adverse consequences from failing to follow through on an agreed-upon sampling or analytic plan.

C6 Strives to make new methodological knowledge widely available to provide benefits to society at large. Presents relevant findings, when possible, to advance public knowledge.

C7 Understands and conforms to confidentiality requirements for data collection, release, and dissemination and any restrictions on its use established by the data provider (to the extent legally required). Protects the use and disclosure of data accordingly. Safeguards privileged information of the employer, client, or funder.

C8 Prioritizes both scientific integrity and the principles outlined in these Guidelines when interests are in conflict.

Internal reporting/communication is as vulnerable to unethical reporting practices as reports for peer or public review. The ethical practitioner has responsibilities to report and communicate fully and completely – including reporting possible options, and deviations from agreed-upon plans- and also to *not report* privileged information. Part of the obligations describe in Principle C is to ensure that what is communicated – clearly – is appropriate, complete, and coherent, so that the reader/hearer/recipient of the report receives the *intended message* of what was communicated. Note that obfuscation of the outputs of statistical practices, particularly when these are intended to mislead or just confuse a stakeholder, conflicts with all obligations to communicate transparently and honestly (throughout the GLs), as well as C2 specifically.

Principle D. Responsibilities to research subjects, data subjects, or those directly affected by statistical practices

The ethical statistical practitioner does not misuse or condone the misuse of data. They protect and respect the rights and interests of human and animal subjects. These responsibilities extend to those who will be directly affected by statistical practices.

The ethical statistical practitioner:

D6 Considers the impact of statistical practice on society, groups, and individuals. Recognizes that statistical practice could adversely affect groups or the public perception of groups, including marginalized groups. Considers approaches to minimize negative impacts in applications or in framing results in reporting.

D7 Refrains from collecting or using more data than is necessary. Uses confidential information only when permitted and only to the extent necessary. Seeks to minimize the risk of re-identification when sharing de-identified data or results where there is an expectation of confidentiality. Explains any impact of de-identification on accuracy of results.

D8   To maximize contributions of data subjects, considers how best to use available data sources for exploration, training, testing, validation, or replication as needed for the application. The ethical statistical practitioner appropriately discloses how the data is used for these purposes and any limitations.

D11 Does not conduct statistical practice that could reasonably be interpreted by subjects as sanctioning a violation of their rights. Seeks to use statistical practices to promote the just and impartial treatment of all individuals.

Research subjects, data subjects, and those directly affected by statistical practices are all stakeholders in effective communication of, or about, statistical practices. Elements D6, D7, D8 and D11 are most specifically relevant for ethical communication and reporting of statistical practices. While D6-D8 are explicitly about communicating, D11 is more implicit. Specifically, the reporting and communicating about the statistical practices must be sufficient to assure that research subjects, data subjects, and those directly affected by the practices can determine that no violations, or sanctioning of violations, have occurred. Recall that Principle C (element C2 in particular) describes the responsibility not to use statistical practices to mislead any stakeholder – D11 reinforces this responsibility.

Principle E. Responsibilities to members of multidisciplinary teams

The ethical statistical practitioner:

E2   Prioritizes these Guidelines for the conduct of statistical practice in cases where ethical guidelines conflict.

E3   Ensures that all communications regarding statistical practices are consistent with these Guidelines. Promotes transparency in all statistical practices.

E4   Avoids compromising validity for expediency. Regardless of pressure on or within the team, does not use inappropriate statistical practices.

Ethical reporting and communication includes considerations of others on the work team- whether or not the team is engaged in research or science. In addition to simply reporting transparently and completely, the ethical practitioner also ensures that what they will be reporting was ethically designed and executed. The ethical obligations of complete and correct reporting and communication are not limited to the final step of practice, but are embedded throughout practice, partly to discharge our Responsibilities to research subjects, data subjects, or those directly affected by statistical practices (D), but also those to our team members (E). Ethical reporting and communicating are both components of ethical statistical practice and also obligations; whenever members of the team seek to suppress or otherwise limit communication about the team's statistical practice or its results, it violates E4, and may also impede the ethical practitioner from following E2.

Principle F. Responsibilities to fellow statistical practitioners and the Profession

The ethical statistical practitioner:

F2 Helps strengthen, and does not undermine, the work of others through appropriate peer review or consultation. Provides feedback or advice that is impartial, constructive, and objective.

F4 Promotes reproducibility and replication, whether results are "significant" or not, by sharing data, methods, and documentation to the extent possible.

F5 Serves as an ambassador for statistical practice by promoting thoughtful choices about data acquisition, analytic procedures, and data structures among non-practitioners and students. Instills appreciation for the concepts and methods of statistical practice.

Ethical reporting is embedded in our responsibilities to other practitioners and to the profession. Ensuring reports that are useable (F4), and encouraging it as fully as possible through constructive peer review or consultation (F2) are two key parts of ethical reporting and communications. F5 more indirectly represents reporting and communication – the ethical statistical practitioner communicates their thoughtful choices, and encourages others to do so as well.

Principle G. Responsibilities of leaders, supervisors, and mentors in statistical practice

Those leading, supervising, or mentoring statistical practitioners are expected to:

G1 Ensure appropriate statistical practice that is consistent with these Guidelines. Protect the statistical practitioners who comply with these Guidelines, and advocate for a culture that supports ethical statistical practice.

G5 Establish a culture that values validation of assumptions, and assessment of model/algorithm performance over time and across relevant subgroups, as needed. Communicate with relevant stakeholders regarding model or algorithm maintenance, failure, or actual or proposed modifications.

The ethical practitioner follows – and supports a culture that prioritizes - the ASA GLs relating to reporting and communication. When ethical reporting is not supported in the workplace, the ethical practitioner recognizes that these behaviors are contrary to the GLs, and that they should not be allowed/followed. Those in leadership roles have a special obligation to promote ethical communication (and following the ASA GLs in general). By encouraging the workplace/employer to uphold, or permit them to uphold, the ASA Ethical GLs for Statistical Practice, the ethical practitioner directly supports the freedom and responsibility of statistical practitioners who comply with these GLs.

Principle H. Responsibilities regarding potential misconduct

The ethical statistical practitioner:

H2 Avoids condoning or appearing to condone statistical, scientific, or professional misconduct. Encourages other practitioners to avoid misconduct or the appearance of misconduct.

The ethical practitioner engages in, and directly and indirectly supports, ethical reporting and communication so as to avoid condoning (and engaging in) incompetent and unprofessional behavior.

ACM Code of Ethics: on reporting your results/ communication

Overall: The Preamble states,

When thinking through a particular issue, a computing professional may find that multiple principles should be taken into account, and that different principles will have different relevance to the issue. Questions related to these kinds of issues can best be answered by thoughtful consideration of the fundamental ethical principles, understanding that the public good is the paramount consideration.

The ethical computing professional has a responsibility to communicate how/that decisions are justifiable and ethical, and/or how the specific features of systems that are designed and deployed have been ethically engineered so as to promote the public good and minimize harms and risks of harms to the extent possible.

1. General Ethical Principles

1.2 A computing professional has an additional obligation to report any signs of system risks that might result in harm. If leaders do not act to curtail or mitigate such risks, it may be necessary to "blow the whistle" to reduce potential harm. However, capricious or misguided reporting of risks can itself be harmful. Before reporting risks, a computing professional should carefully assess relevant aspects of the situation.

1.3 A computing professional should be transparent and provide full disclosure of all pertinent system capabilities, limitations, and potential problems to the appropriate parties.

1.6 Computing professionals should establish transparent policies and procedures that allow individuals to understand what data is being collected and how it is being used, to give informed consent for automatic data collection, and to review, obtain, correct inaccuracies in, and delete their personal data.

Only the minimum amount of personal information necessary should be collected in a system. The retention and disposal periods for that information should be clearly defined, enforced, and communicated to data subjects. Personal information gathered for a specific purpose should not be used for other purposes without the person's consent. Merged data collections can compromise privacy features present in the original collections. Therefore, computing professionals should take special care for privacy when merging data collections.

ACM Principle 1 includes specifications of responsibility for clear and honest communication with relevant stakeholders. As discussed under data collection (Chapter 2.3), the ethical practitioner transparently reports and communicates to data contributors exactly what data is being collected, otherwise, consent for data collection cannot be characterized as "informed". The computing professional must therefore communicate what the data collection system is doing, and for what purposes to the potential data contributor.

Since the ACM articulates a specific responsibility by all ethical practitioners to protect individual rights and the public good, ACM Principle 1.1 (among others) points to an obligation to communicate honestly how harms or benefits may (or do) accrue to the individual (and more generally, to the public).

2. Professional Responsibilities

2.1 The dignity of employers, employees, colleagues, clients, users, and anyone else affected either directly or indirectly by the work should be respected throughout the process. Computing professionals should respect the right of those involved to transparent communication about the project. Professionals should be cognizant of any serious negative consequences affecting any stakeholder that may result from poor quality work and should resist inducements to neglect this responsibility.

2.4 Computing professionals should also provide constructive, critical reviews of others' work.

2.5 Computing professionals are in a position of trust, and therefore have a special responsibility to provide objective, credible evaluations and testimony to employers, employees, clients, users, and the public. Computing professionals should strive to be perceptive, thorough, and objective when evaluating, recommending, and presenting system descriptions and alternatives. Extraordinary care should be taken to identify and mitigate potential risks in machine learning systems. A system for which future risks cannot be reliably predicted requires frequent reassessment of risk as the system evolves in use, or it should not be deployed. Any issues that might result in major risk must be reported to appropriate parties.

2.6 A computing professional is responsible for evaluating potential work assignments. This includes evaluating the work's feasibility and advisability, and making a judgment about whether the work assignment is within the professional's areas of competence. If at any time before or during the work assignment the professional identifies a lack of a necessary expertise, they must disclose this to the employer or client.

2.9 Computing professionals should perform due diligence to ensure the system functions as intended, and take appropriate action to secure resources against accidental and intentional misuse, modification, and denial of service. As threats can arise and change after a system is deployed, computing professionals should integrate mitigation techniques and policies, such as monitoring, patching, and vulnerability reporting.

Clear and effective communication is required to disclose, or report, all aspects of system capabilities, limitations, and their potential for harms. "Objective, credible" evaluations of systems represent effective communication, discharging the Professional Responsibilities under ACM Principle 2.

3. Professional Leadership Principles

3.1 People—including users, customers, colleagues, and others affected directly or indirectly— should always be the central concern in computing. The public good should always be an explicit consideration when evaluating tasks associated with research, requirements analysis, design, implementation, testing, validation, deployment, maintenance, retirement, and disposal. Computing professionals should keep this focus no matter which methodologies or techniques they use in their practice.

3.4 Leaders should pursue clearly defined organizational policies that are consistent with the Code and effectively communicate them to relevant stakeholders. In addition, leaders should encourage and reward compliance with those policies, and take appropriate action when policies are violated. Designing or implementing processes that deliberately or negligently violate, or tend to enable the violation of, the Code's principles is ethically unacceptable.

3.5 Computing professionals should be fully aware of the dangers of oversimplified approaches, the improbability of anticipating every possible operating condition, the inevitability of software errors, the interactions of systems and their contexts, and other issues related to the complexity of their profession—and thus be confident in taking on responsibilities for the work that they do.

Reiterating the emphasis on the public good and avoiding harm, the ethical computing professional must both competently evaluate and design systems (or assignments to create systems), and also communicate the capabilities, limitations, and potential (or actual) harms of these systems to relevant stakeholders. Moreover, leaders in computing have an additional obligation to ensure that policies exist, or are created, that are communicated sufficiently to all stakeholders. It is impossible for leaders to support employees, colleagues, and clients following the ACM CE if policies that enable/ensure the CEs are followed are not communicated effectively.

4. Compliance with the Code

4.1 Computing professionals who recognize breaches of the Code should take actions to resolve the ethical issues they recognize, including, when reasonable, expressing their concern to the person or persons thought to be violating the Code.

4.2 ACM members who recognize a breach of the Code should consider reporting the violation.

As discussed for other ACM Principles, the identification of failures to comply with the ACM CE will require explicit communication – to the person thought to be violating the code, and to the ACM itself. Enumerating CE principles violated, or simple notification that "someone isn't following the CE" will clearly be insufficient, so consideration of how results of the computing professional's reasoning were obtained, as well as those results, must be communicated clearly.

The stakeholder analysis (SHA) table below shows some harms, benefits, unknown outcomes, and unknowable outcomes, relating to following the GLs and CE in reporting and communication.

Potential result: Stakeholder:	HARM	BENEFIT	UNKNOWN	UNKNOWABLE
YOU	Takes time, adds effort (to fully document for specific reports/ audiences); may facilitate use of work that is unauthorized/ does not acknowledge original creator	Simplifies evaluations of the work (ASA) or system, and updates (ACM); adds transparency and accountability.	ACM: New harms/risks may be detectable more quickly (and addressed) with reports that communicate effectively. ASA: replications are facilitated.	
Your boss/client	Takes time	Creates transparency and accountability of statistician/ computing professional work		
Unknown individuals	Accessible (audience-appropriate) communication may alert stakeholders to potential harms of which they would otherwise be ignorant. This could lead to them complaining, or withdrawing /withholding consent	Creates transparency and accountability		

Potential result: Stakeholder:	HARM	BENEFIT	UNKNOWN	UNKNOWABLE
Employer	Takes time	Demonstrates commitment to transparency & accountability.	ACM: New harms/risks may be detectable more quickly (and addressed) with full documentation. ASA: replications are facilitated.	Accessible (audience-appropriate) communication may alert stakeholders to potential harms of which they would otherwise be ignorant. This could lead to them complaining, or withdrawing/withholding consent
Colleagues	May add time (in reviewing)	Supports transparency, accountability, and replicability. Promotes peer evaluation as well as wider audience comprehension of your work.		
Profession	Takes time, adds effort (to fully document plans/evaluations/ systems); may facilitate use of work that is unauthorized/ does not acknowledge original creator	Simplifies evaluations of the work (ASA) or system, and updates (ACM); adds transparency and accountability. Promotes and facilitates peer evaluation.	ACM: encourages careful review, so new harms/risks may be detectable more quickly (and addressed). ASA: transparency and validity of analyses and interpretations are facilitated.	

Potential result:	HARM	BENEFIT	UNKNOWN	UNKNOWABLE
Stakeholder:				
Public/public trust	Perpetuating black box mentality does not help, and explicitly does not inform, society. Accessible (audience-appropriate) communication may alert stakeholders to potential harms of which they would otherwise be ignorant. This could lead to them complaining, or withdrawing /withholding consent.	Transparency and accountability – and the potential to understand how decisions are made based on automation and systems - supports public trust in the profession and its work.	Transparency supports trust, even if errors are discovered; their discoverability arises only and specifically from documentation, transparency, and accountability.	

Table 2.7.1. *Stakeholder Analysis template: Reporting your results/communication*

Questions for Discussion:

Discuss similarities and differences between the ASA GLs and ACM CE principles that are discussed above, in terms of how they apply to communicating/reporting results at work. Does either practice standard suggest that communication or reporting should differ for the workplace or for/to the public?

What are the most striking similarities (e.g., what is unexpectedly similar)? Are there particular differences that stand out because they are A) highly predictable? B) unexpected? C) have different benefits or harms to different stakeholders?

Rank GL/CE elements from most to least important in terms of reporting your results/communication. What does the rankability mean for you or for the discipline of statistics and data science?

Can following the GL/CE affect how you communicate/report results in your typical work? How/How not?

Is the SHA in Table 2.7.1 enough to ensure ethical practice? Why/why not?

Discuss how the ASA and ACM are, or are not, fully aligned with "responsible authorship and publication".

Discuss how the ASA and ACM are, or are not, fully aligned with current policies (for your institution, organization, or lab/work team) on "conflicts of interest".

Chapter 2.8
Engaging in team science/teamwork

ASA Ethical Guidelines (GLs) for Statistical Practice: on engaging in team science/teamwork

Overall: The preamble to the GLs states that:

In these Guidelines, "statistical practice" includes activities such as: designing the collection of, summarizing, processing, analyzing, interpreting, or presenting, data; as well as model or algorithm development and deployment. Throughout these Guidelines, the term "statistical practitioner" includes *all those who engage in statistical practice, regardless of job title, profession, level, or field of degree.* The Guidelines are intended for individuals, but these principles are also relevant to organizations that engage in statistical practice. All statistical practitioners are expected to follow these Guidelines and to encourage others to do the same. (Emphasis added.)

The ethical practitioner strives for ethical "statistical practice" as much for their own professionalism as to support the work of those around them and particularly, the decisions that may be based on their work. In team science and any teamwork, all statistical practice should be transparent, reproducible, and valid. Sometimes the statistical practitioner completes their work alone and sometimes works with a team of other statisticians, but the ethical practitioner always serves their team(s), their profession, and their own reputation by contributing to science or other evidence-based work and promoting ethical work by others.

Principle A: Professional integrity and accountability

Professional integrity and accountability require taking responsibility for one's work. Ethical statistical practice supports valid and prudent decision making with appropriate methodology. The ethical statistical practitioner represents their capabilities and activities honestly, and treats others with respect.

The ethical statistical practitioner:

A1 Takes responsibility for evaluating potential tasks, assessing whether they have (or can attain) sufficient competence to execute each task, and that the work and timeline are feasible. Does not solicit or deliver work for which they are not qualified, or that they would not be willing to have peer reviewed.

A2 Uses methodology and data that are valid, relevant, and appropriate, without favoritism or prejudice, and in a manner intended to produce valid, interpretable, and reproducible results.

A3 Does not knowingly conduct statistical practices that exploit vulnerable populations or create or perpetuate unfair outcomes.

A4 Opposes efforts to predetermine or influence the results of statistical practices, and resists pressure to selectively interpret data.

A5 Accepts full responsibility for their own work; does not take credit for the work of others; and gives credit to those who contribute. Respects and acknowledges the intellectual property of others.

A6 Strives to follow, and encourages all collaborators to follow, an established protocol for authorship. Advocates for recognition commensurate with each person's contribution to the work. Recognizes that inclusion as an author does imply, while acknowledgement may imply, endorsement of the work.

A7 Discloses conflicts of interest, financial and otherwise, and manages or resolves them according to established policies, regulations, and laws.

A8 Promotes the dignity and fair treatment of all people. Neither engages in nor condones discrimination based on personal characteristics. Respects personal boundaries in interactions and avoids harassment including sexual harassment, bullying, and other abuses of power or authority.

A9 Takes appropriate action when aware of deviations from these Guidelines by others.

A10 Acquires and maintains competence through upgrading of skills as needed to maintain a high standard of practice.

A11 Follows applicable policies, regulations, and laws relating to their professional work, unless there is a compelling ethical justification to do otherwise.

A12 Upholds, respects, and promotes these Guidelines. Those who teach, train, or mentor in statistical practice have a special obligation to promote behavior that is consistent with these Guidelines.

The entirety of Principle A – relating to professional integrity and accountability – is relevant for the ethical practitioner engaging in team science or work. Fulfilling Principle A requires competence in the domain, honest self-representation, encouragement for all team members to recognize and accept their responsibilities as statistical practitioners, and respectful treatment of, and communication with, others.

Principle B. Integrity of data and methods

The ethical statistical practitioner seeks to understand and mitigate known or suspected limitations, defects, or biases in the data or methods and communicates potential impacts on the interpretation, conclusions, recommendations, decisions, or other results of statistical practices.

The ethical statistical practitioner:

B1 Communicates data sources and fitness for use, including data generation and collection processes and known biases. Discloses and manages any conflicts of interest relating to the data sources. Communicates data processing and transformation procedures, including missing data handling.

B2 Is transparent about assumptions made in the execution and interpretation of statistical practices including methods used, limitations, possible sources of error, and algorithmic biases. Conveys results or applications of statistical practices in ways that are honest and meaningful.

B3 Communicates the stated purpose and the intended use of statistical practices. Is transparent regarding a priori versus post hoc objectives and planned versus unplanned statistical practices. Discloses when multiple comparisons are conducted, and any relevant adjustments.

B4 Meets obligations to share the data used in the statistical practices, for example, for peer review and replication, as allowable. Respects expectations of data contributors when using or sharing data. Exercises due caution to protect proprietary and confidential data, including all data that might inappropriately harm data subjects.

B5 Strives to promptly correct substantive errors discovered after publication or implementation. As appropriate, disseminates the correction publicly and/or to others relying on the results.

B6 For models and algorithms designed to inform or implement decisions repeatedly, develops and/or implements plans to validate assumptions and assess performance over time, as needed. Considers criteria and mitigation plans for model or algorithm failure and retirement.

B7   Explores and describes the effect of variation in human characteristics and groups on statistical practice when feasible and relevant.

Every element in Principle B relates specifically and/or indirectly to ethical statistical work as a part of a team (as well as outside of a team!). Critically, the ethical statistical practitioner may be the only team member who fully understands the importance of limitations/defects or bias in the data, and may be solely responsible for the integrity of methods used to address the team goal. The ethical statistical practitioner must communicate clearly and effectively to ensure that all team members understand and respect the integrity and provenance of the data, limitations of the data and methods, and the role of uncertainty in all results.

Principle C. Responsibilities to stakeholders

Those who fund, contribute to, use, or are affected by statistical practices are considered stakeholders. The ethical statistical practitioner respects the interests of stakeholders while practicing in compliance with these Guidelines.

The ethical statistical practitioner:

C1   Seeks to establish what stakeholders hope to obtain from any specific project. Strives to obtain sufficient subject-matter knowledge to conduct meaningful and relevant statistical practice.

C2   Regardless of personal or institutional interests or external pressures, does not use statistical practices to mislead any stakeholder.

C3   Uses practices appropriate to exploratory and confirmatory phases of a project, differentiating findings from each so the stakeholders can understand and apply the results.

C4   Informs stakeholders of the potential limitations on use and re-use of statistical practices in different contexts and offers guidance and alternatives, where appropriate, about scope, cost, and precision considerations that affect the utility of the statistical practice.

C5   Explains any expected adverse consequences from failing to follow through on an agreed-upon sampling or analytic plan.

C6   Strives to make new methodological knowledge widely available to provide benefits to society at large. Presents relevant findings, when possible, to advance public knowledge.

C7   Understands and conforms to confidentiality requirements for data collection, release, and dissemination and any restrictions on its use established by the data provider (to the extent legally required). Protects the use and disclosure of data accordingly. Safeguards privileged information of the employer, client, or funder.

C8   Prioritizes both scientific integrity and the principles outlined in these Guidelines when interests are in conflict.

The statistician must support their team and its objectives while keeping in mind their specific responsibilities to other stakeholders, including the profession. In cases where team benefits come at the cost of harms to science (or the scientific record), or the public, these must be carefully contemplated, and the resolution of such a conflict should be informed by the responsibilities of all team members to the relevant stakeholders, and not solely by team goals. Avoiding misleading any stakeholder is the principal objective of Principle C's elements. It might not be sufficient to present or document the statistical work ethically, and then hope that others on the team recognize when benefits might accrue to the team at the cost of harms to stakeholders and act to mitigate those harms. Thus, the ethical statistical practitioner may need to follow up on recommendations, as it is possible, to ensure that undue harms are not the result of hoped-for, but not observed, prioritization of key stakeholders.

Principle D. Responsibilities to research subjects, data subjects, or those directly affected by statistical practices

The ethical statistical practitioner does not misuse or condone the misuse of data. They protect and respect the rights and interests of human and animal subjects. These responsibilities extend to those who will be directly affected by statistical practices.

The ethical statistical practitioner:

D1   Keeps informed about and adheres to applicable rules, approvals, and guidelines for the protection and welfare of human and animal subjects. Knows when work requires ethical review and oversight.

D2   Makes informed recommendations for sample size and statistical practice methodology in order to avoid the use of excessive or inadequate numbers of subjects and excessive risk to subjects.

D3   For animal studies, seeks to leverage statistical practice to reduce the number of animals used, refine experiments to increase the humane treatment of animals, and replace animal use where possible.

D4   Protects people's privacy and the confidentiality of data concerning them, whether obtained from the individuals directly, other persons, or existing records. Knows and adheres to applicable rules, consents, and guidelines to protect private information.

D5   Uses data only as permitted by data subjects' consent when applicable or considering their interests and welfare when consent is not required. This includes primary and secondary uses, use of repurposed data, sharing data, and linking data with additional data sets.

D6   Considers the impact of statistical practice on society, groups, and individuals. Recognizes that statistical practice could adversely affect groups or the public perception of groups, including marginalized groups. Considers approaches to minimize negative impacts in applications or in framing results in reporting.

D7   Refrains from collecting or using more data than is necessary. Uses confidential information only when permitted and only to the extent necessary. Seeks to minimize the risk of re-identification when sharing de-identified data or results where there is an expectation of confidentiality. Explains any impact of de-identification on accuracy of results.

D8   To maximize contributions of data subjects, considers how best to use available data sources for exploration, training, testing, validation, or replication as needed for the application. The ethical statistical practitioner appropriately discloses how the data is used for these purposes and any limitations.

D9   Knows the legal limitations on privacy and confidentiality assurances and does not over-promise or assume legal privacy and confidentiality protections where they may not apply.

D10   Understands the provenance of the data, including origins, revisions, and any restrictions on usage, and fitness for use prior to conducting statistical practices.

D11   Does not conduct statistical practice that could reasonably be interpreted by subjects as sanctioning a violation of their rights. Seeks to use statistical practices to promote the just and impartial treatment of all individuals.

Sometimes the statistical practitioner is the only person on a team who sees/works with the data; and some teams do not approach a statistical practitioner until after they have collected or obtained data. The ethical statistical practitioner must consider whether or not to join a team or project given their Responsibilities to research subjects, data subjects, or those directly affected by statistical practices (human or animal), and even if they joined a team prior to data collection, if "participating in the analysis could reasonably be interpreted by individuals who provided information as sanctioning a violation of their rights", then the statistician should consider leaving the team. In addition to avoiding misleading any stakeholder (C2), a key responsibility of the ethical practitioner is to know data provenance (D10), ensure data are fit for the team's purposes (D10), and ensure that subjects would not interpret the team's work as a violation or sanction of violation of their rights (D11). These responsibilities exist even if data subjects will never know the results of the statistical practice.

Principle E. Responsibilities to members of multidisciplinary teams

The ethical statistical practitioner:

E1 Recognizes and respects that other professions may have different ethical standards and obligations. Dissonance in ethics may still arise even if all members feel that they are working towards the same goal. It is essential to have a respectful exchange of views.

E2 Prioritizes these Guidelines for the conduct of statistical practice in cases where ethical guidelines conflict.

E3 Ensures that all communications regarding statistical practices are consistent with these Guidelines. Promotes transparency in all statistical practices.

E4 Avoids compromising validity for expediency. Regardless of pressure on or within the team, does not use inappropriate statistical practices

The ethical practitioner has an obligation –to the scientific community, the public, and other stakeholders (C), to the research and data subjects and those affected by statistical practice (D) - to ensure that what the team will be doing and reporting was ethically designed and executed. The statistical practitioner's responsibilities to team colleagues (E) emphasize appropriate methods, transparency, and following these Guidelines. Thus, if team colleague practice standards, or the team's priorities, are in conflict with the ASA GLs, then the ethical practitioner prioritizes their responsibility to follow these Guidelines. This can be challenging, particularly given the respectful exchange of views required in E1. However, for a team engaged in statistical practice as defined in the Purpose of the GLs, all work relating to statistical practices – irrespective of other interests – fall under these GLs.

Principle F. Responsibilities to fellow statistical practitioners and the Profession

The ethical statistical practitioner:

F1 Recognizes that statistical practitioners may have different expertise and experiences, which may lead to divergent judgments about statistical practices and results. Constructive discourse with mutual respect focuses on scientific principles and methodology and not personal attributes.

F2 Helps strengthen, and does not undermine, the work of others through appropriate peer review or consultation. Provides feedback or advice that is impartial, constructive, and objective.

F3 Takes full responsibility for their contributions as instructors, mentors, and supervisors of statistical practice by ensuring their best teaching and advising -- regardless of an academic or non-academic setting -- to ensure that developing practitioners are guided effectively as they learn and grow in their careers.

F4 Promotes reproducibility and replication, whether results are "significant" or not, by sharing data, methods, and documentation to the extent possible.

F5 Serves as an ambassador for statistical practice by promoting thoughtful choices about data acquisition, analytic procedures, and data structures among non-practitioners and students. Instills appreciation for the concepts and methods of statistical practice.

The entirety of Principle F is relevant for working on teams, even if there is only one statistical practitioner. Consideration of how others in the profession would consider the team's work can be informative about whether the responsibilities listed in Principle F have been met. If the practitioner fears that another statistical practitioner would feel uncomfortable with how a team's work is planned, done, documented, or communicated, then it suggests that Principle F has not been fulfilled. When team-based work is not solely focused on statistical practice, respect for others is the paramount consideration in all dealings with statistical practitioners on or off the team, however the ethical practitioner is an ambassador for ethical statistical practice – especially on teams. Not all statistics practitioners are aware of the ASA GLs, so even on teams with more than one statistical practitioner, the role of ambassador is essential.

Principle G. Responsibilities of leaders, supervisors, and mentors in statistical practice

Those leading, supervising, or mentoring statistical practitioners are expected to:

G1 Ensure appropriate statistical practice that is consistent with these Guidelines. Protect the statistical practitioners who comply with these Guidelines, and advocate for a culture that supports ethical statistical practice.

G2 Promote a respectful, safe, and productive work environment. Encourage constructive engagement to improve statistical practice.

G3 Identify and/or create opportunities for team members/mentees to develop professionally and maintain their proficiency.

G4 Advocate for appropriate, timely, inclusion and participation of statistical practitioners as contributors/collaborators. Promote appropriate recognition of the contributions of statistical practitioners, including authorship if applicable.

G5 Establish a culture that values validation of assumptions, and assessment of model/algorithm performance over time and across relevant subgroups, as needed. Communicate with relevant stakeholders regarding model or algorithm maintenance, failure, or actual or proposed modifications.

The ethical practitioner follows the ASA GLs, even when on teams where others seek to follow other practice standards. When the workplace (or the team) does not support the statistician following the GLs, or if violating ASL GL principles is encouraged on teams (or off of them), the ethical practitioner recognizes that these behaviors are contrary to the GLs, and that they should not be allowed/followed. The ethical practitioner supports the profession as well as the freedom and responsibility of statistical practitioners who comply with these GLs, whether they are on their team or not. It may come to pass that an ethical statistician seeks a different job/workplace or team to avoid team or workplace pressures to violate ASA Ethical Guidelines for Statistical Practice.

Principle H. Responsibilities regarding potential misconduct

The ethical statistical practitioner understands that questions may arise concerning potential misconduct related to statistical, scientific, or professional practice. At times, a practitioner may accuse someone of misconduct, or be accused by others. At other times, a practitioner may be involved in the investigation of others' behavior. Allegations of misconduct may arise within different institutions with different standards and potentially different outcomes. The elements that follow relate specifically to allegations of statistical, scientific, and professional misconduct.

The ethical statistical practitioner:

H1 Knows the definitions of, and procedures relating to, misconduct in their institutional setting. Seeks to clarify facts and intent before alleging misconduct by others. Recognizes that differences of opinion and honest error do not constitute unethical behavior.

H2 Avoids condoning or appearing to condone statistical, scientific, or professional misconduct. Encourages other practitioners to avoid misconduct or the appearance of misconduct.

H3 Does not make allegations that are poorly founded, or intended to intimidate. Recognizes such allegations as potential ethics violations.

H4 Lodges complaints of misconduct discreetly and to the relevant institutional body. Does not act on allegations of misconduct without appropriate institutional referral, including those allegations originating from social media accounts or email listservs.

H5 Insists upon a transparent and fair process to adjudicate claims of misconduct. Maintains confidentiality when participating in an investigation. Discloses the investigation results honestly to appropriate parties and stakeholders once they are available.

H6 Refuses to publicly question or discredit the reputation of a person based on a specific accusation of misconduct while due process continues to unfold.

H7 Following an investigation of misconduct, supports the efforts of all parties involved to resume their careers in as normal a manner as possible, consistent with the outcome of the investigation.

H8 Avoids, and acts to discourage, retaliation against or damage to the employability of those who responsibly call attention to possible misconduct.

The ethical practitioner engages in, and directly and indirectly supports, ethical and professional conduct (scientific, statistical, and professional). Part of the rationale for adhering, and supporting/encouraging team members to adhere, to these GLs is because ethical statistical practice represents the *opposite* of statistical, scientific, and professional **misconduct**. All team science and teamwork should avoid both condoning and engaging in incompetent and unprofessional behavior, so the ethical statistical practitioner follows the GLs whether or not they are part of a team. Intentionally misleading a stakeholder or future user of whatever work the statistical practitioner and their team engages in undermines the statistical profession as well as demonstrating a lack of integrity, and is unprofessional. However, investigations into misconduct may involve team members and the ethical statistical practitioner must follow all of the Principle H elements in those cases. However, the elements of Principle H are relevant whenever there is potential misconduct, not only when the statistical practitioner is on a team.

ACM Code of Ethics: on Engaging in team science/ teamwork

Overall: The Preamble states,

When thinking through a particular issue, a computing professional may find that multiple principles should be taken into account, and that different principles will have different relevance to the issue. Questions related to these kinds of issues can best be answered by thoughtful consideration of the fundamental ethical principles, understanding that the public good is the paramount consideration. The entire computing profession benefits when the ethical decision-making process is accountable to and transparent to all stakeholders. Open discussions about ethical issues promote this accountability and transparency.

The ethical computing professional has a responsibility to do work that supports the public good, and to ensure that decisions in all aspects of computing are justifiable and ethical. Teams are an excellent opportunity for the open discussions that promote accountability and transparency, so that the result of team science/work promotes the public good and minimizes harms and risks of harms. The computing professional keeps the public good as their primary focus, prioritizing this over other stakeholders when necessary, which might happen in team-based work.

1. General Ethical Principles

1.1 Contribute to society and to human well-being, acknowledging that all people are stakeholders in computing.
     This principle, which concerns the quality of life of all people, affirms an obligation of computing professionals, both individually and collectively, to use their skills for the benefit of society, its members, and the environment surrounding them. This obligation includes promoting fundamental human rights and protecting each individual's right to autonomy. An essential aim of computing professionals is to minimize negative consequences of computing,

including threats to health, safety, personal security, and privacy. When the interests of multiple groups conflict, the needs of those less advantaged should be given increased attention and priority.

Computing professionals should consider whether the results of their efforts will respect diversity, will be used in socially responsible ways, will meet social needs, and will be broadly accessible.

1.2 voiding harm begins with careful consideration of potential impacts on all those affected by decisions. When harm is an intentional part of the system, those responsible are obligated to ensure that the harm is ethically justified. In either case, ensure that all harm is minimized.

To minimize the possibility of indirectly or unintentionally harming others, computing professionals should follow generally accepted best practices unless there is a compelling ethical reason to do otherwise. Additionally, the consequences of data aggregation and emergent properties of systems should be carefully analyzed.

A computing professional has an additional obligation to report any signs of system risks that might result in harm. If leaders do not act to curtail or mitigate such risks, it may be necessary to "blow the whistle" to reduce potential harm. However, capricious or misguided reporting of risks can itself be harmful. Before reporting risks, a computing professional should carefully assess relevant aspects of the situation.

1.3 A computing professional should be transparent and provide full disclosure of all pertinent system capabilities, limitations, and potential problems to the appropriate parties. Making deliberately false or misleading claims, fabricating or falsifying data, offering or accepting bribes, and other dishonest conduct are violations of the Code.

1.4 The use of information and technology may cause new, or enhance existing, inequities.

Computing professionals should foster fair participation of all people, including those of underrepresented groups. Prejudicial discrimination on the basis of age, color, disability, ethnicity, family status, gender identity, labor union membership, military status, nationality, race, religion or belief, sex, sexual orientation, or any other inappropriate factor is an explicit violation of the Code. Harassment, including sexual harassment, bullying, and other abuses of power and authority, is a form of discrimination that, amongst other harms, limits fair access to the virtual and physical spaces where such harassment takes place.

Technologies and practices should be as inclusive and accessible as possible and computing professionals should take action to avoid creating systems or technologies that disenfranchise or oppress people.

1.5 Computing professionals should therefore credit the creators of ideas, inventions, work, and artifacts, and respect copyrights, patents, trade secrets, license agreements, and other methods of protecting authors' works.

1.6 Computing professionals should only use personal information for legitimate ends and without violating the rights of individuals and groups. This requires taking precautions to prevent re- identification of anonymized data or unauthorized data collection, ensuring the accuracy of data, understanding the provenance of the data, and protecting it from unauthorized access and accidental disclosure.

Computing professionals should establish transparent policies and procedures that allow individuals to understand what data is being collected and how it is being used, to give informed consent for automatic data collection, and to review, obtain, correct inaccuracies in, and delete their personal data.

Only the minimum amount of personal information necessary should be collected in a system. The retention and disposal periods for that information should be clearly defined, enforced, and communicated to data subjects. Personal information gathered for a specific purpose should not be used for other purposes without the person's consent. Merged data collections can compromise privacy features present in the original collections. Therefore, computing professionals should take special care for privacy when merging data collections.

1.7 Computing professionals should protect confidentiality except in cases where it is evidence of the violation of law, of organizational regulations, or of the Code. In these cases, the nature or contents of that information should not be disclosed except to appropriate authorities. A computing professional should consider thoughtfully whether such disclosures are consistent with the Code.

All of ACM Principle 1 relates directly or indirectly to how the ethical computing professional will perform their work, whether on teams or not. Particularly when working on teams, each computing professional has an obligation to "use their skills for the benefit of society, its members, and the environment surrounding them" (1.1), and in a team context, this may refer to computing as well as communication skills. The computing professional has opportunities to encourage other team members, irrespective of their ACM membership, to follow the ACM CE.

2. Professional Responsibilities

2.1   Computing professionals should insist on and support high quality work from themselves and from colleagues. The dignity of employers, employees, colleagues, clients, users, and anyone else affected either directly or indirectly by the work should be respected throughout the process. Computing professionals should respect the right of those involved to transparent communication about the project. Professionals should be cognizant of any serious negative consequences affecting any stakeholder that may result from poor quality work and should resist inducements to neglect this responsibility.

2.2   High quality computing depends on individuals and teams who take personal and group responsibility for acquiring and maintaining professional competence. Professional competence starts with technical knowledge and with awareness of the social context in which their work may be deployed. Professional competence also requires skill in communication, in reflective analysis, and in recognizing and navigating ethical challenges. Upgrading skills should be an ongoing process and might include independent study, attending conferences or seminars, and other informal or formal education. Professional organizations and employers should encourage and facilitate these activities.

2.3   Rules that are judged unethical should be challenged. A rule may be unethical when it has an inadequate moral basis or causes recognizable harm.

2.4   Computing professionals should also provide constructive, critical reviews of others' work.

2.5   Computing professionals are in a position of trust, and therefore have a special responsibility to provide objective, credible evaluations and testimony to employers, employees, clients, users, and the public. Computing professionals should strive to be perceptive, thorough, and objective when evaluating, recommending, and presenting system descriptions and alternatives. Extraordinary care should be taken to identify and mitigate potential risks in machine learning systems. A system for which future risks cannot be reliably predicted requires frequent reassessment of risk as the system evolves in use, or it should not be deployed. Any issues that might result in major risk must be reported to appropriate parties.

2.6   A computing professional is responsible for evaluating potential work assignments. This includes evaluating the work's feasibility and advisability, and making a judgment about whether the work assignment is within the professional's areas of competence. If at any time before or during the work assignment the professional identifies a lack of a necessary expertise, they must disclose this to the employer or client.

2.8 Under exceptional circumstances a computing professional may use unauthorized access to disrupt or inhibit the functioning of malicious systems; extraordinary precautions must be taken in these instances to avoid harm to others.

2.9 Computing professionals should perform due diligence to ensure the system functions as intended, and take appropriate action to secure resources against accidental and intentional misuse, modification, and denial of service. As threats can arise and change after a system is deployed, computing professionals should integrate mitigation techniques and policies, such as monitoring, patching, and vulnerability reporting.

In cases where misuse or harm are predictable or unavoidable, the best option may be to not implement the system.

The professional responsibilities (Principle 2) of ethical computing involve extensive contextual and stakeholder analysis, and these pertain to work done on teams or alone. "High quality computing depends on individuals and teams who take personal and group responsibility for acquiring and maintaining professional competence." In this (2.2) is an echo of Principle 4 (compliance with the CE), namely that each individual on a team must take responsibility for ensuring that the team's work is done ethically and competently. Similarly, ACM 2.2 identifies competencies beyond computing expertise as essential professional responsibilities: "Professional competence also requires skill in communication, in reflective analysis, and in recognizing and navigating ethical challenges." Communication as well as competence are required to ensure that computing work, reviews, evaluations, and stakeholder analyses are all carefully done and delivered to team members in a manner that supports the other team members' appreciation – and eventual use – of the input.

3. Professional Leadership Principles

3.1 Ensure that the public good is the central concern during all professional computing work.

People—including users, customers, colleagues, and others affected directly or indirectly— should always be the central concern in computing. The public good should always be an explicit consideration when evaluating tasks associated with research, requirements analysis, design, implementation, testing, validation, deployment, maintenance, retirement, and disposal. Computing professionals should keep this focus no matter which methodologies or techniques they use in their practice.

3.4 Articulate, apply, and support policies and processes that reflect the principles of the Code.

Leaders should pursue clearly defined organizational policies that are consistent with the Code and effectively communicate them to relevant stakeholders. In addition, leaders should encourage and reward compliance with those policies, and take appropriate action when policies are violated. Designing or implementing processes that deliberately or negligently violate, or tend to enable the violation of, the Code's principles is ethically unacceptable.

3.5 Create opportunities for members of the organization or group to grow as professionals.

Computing professionals should be fully aware of the dangers of oversimplified approaches, the improbability of anticipating every possible operating condition, the inevitability of software errors, the interactions of systems and their contexts, and other issues related to the complexity of their profession—and thus be confident in taking on responsibilities for the work that they do.

Reiterating the emphasis on the public good and avoiding harm, the ethical computing professional on teams may lead in terms of ensuring ethical decision making; leaders of computing projects or teams have a special obligation to "Ensure that the public good is the central concern during all professional computing work", which may be effected through leadership of the team's work (3.1) or by ensuring that policies are in place and utilized (3.4) to promote ethical decision making on all teams and work. A special obligation of leaders in computing is to promote the professional development of their team members; since ACM 2.2 articulates that "Professional competence also requires skill in communication, in reflective analysis, and in recognizing and navigating ethical challenges", leaders can strengthen both their team members' professionalism and the entire team's commitment to prioritizing the public and minimizing harms by engaging with the ACM CE.

4. Compliance with the Code

4.1   Uphold, promote, and respect the principles of the Code.

Computing professionals who recognize breaches of the Code should take actions to resolve the ethical issues they recognize, including, when reasonable, expressing their concern to the person or persons thought to be violating the Code.

4.2 Each ACM member should encourage and support adherence by all computing professionals regardless of ACM membership.

An important aspect of the ACM CE is its focus on ensuring that all involved in computing, whether or not they are members of the ACM, should be encouraged to respect the CE and its principles. The ethical computing professional should be able to identify failures to comply with the ACM CE by team members ("regardless of ACM membership") and will require the multiple professional competencies in "communication, in reflective analysis, and in recognizing and navigating ethical challenges" so that ACM 4 can be followed by all team members (and all computing professionals!).

The stakeholder analysis (SHA) table below shows some harms, benefits, unknown outcomes, and unknowable outcomes, relating to following the GLs and CE in working on a team.

Potential result: Stakeholder:	HARM	BENEFIT	UNKNOWN	UNKNOWABLE
YOU	Following GL/CE takes extra time, adds effort, and may require confrontations.	Compliance with CE/GL strengthens individual professionalism, the team's work, and the profession as well.	Following GL/CE helps protects against violations of others' rights and promotes informedness of society & decisions people make.	Failures to follow GL/CE can lead to unpredictable behavior of teams/team members. Failures can lead to variable, not high quality, work.
Your boss/client	Following GL/CE takes extra time, adds effort, and may result in confrontations between ACM/ASA members and others on teams. Following GL/CE may cost time/money and may not yield desired results.	ACM CE prioritizes public good and quality of life of all people, while ASA GL prioritize transparency, reproducibility, and validity of work. These are benefits to all.	Reputation for following CE/GL may drive some clients away while attracting others.	

Potential result: Stakeholder:	HARM	BENEFIT	UNKNOWN	UNKNOWABLE
Unknown individuals	Failures to follow GL/CE can lead to violated rights and acts against social/public good. Irreproducible results and low-quality work lower, and do not augment, informedness. Failures can increase risks of bias and harms to individuals, science, and the public.	GL/CE represent best practices for protecting rights and promoting social/public good and informedness. GL/CE mitigate risks of bias and harms.	Even following GL/CE explicitly cannot avoid all possible harms. However, failing to follow GL/CE can mean not trying to avoid or minimize these harms at all.	
Employer	Following GL/CE takes extra time, adds effort, and may result in confrontations between ACM/ASA members and others on teams. Following GL/CE may cost time/money and may not yield desired results.	ACM CE prioritizes public good and quality of life of all people, while ASA GL prioritize transparency, reproducibility, and validity of work. These are benefits to all.	Reputation for following CE/GL may drive some clients away while attracting others.	Could limit business/profit if other practitioners/ organizations can be found that do not follow CE/GL.

Potential result: Stakeholder:	HARM	BENEFIT	UNKNOWN	UNKNOWABLE
Colleagues	May complicate – or undermine -colleagues' work (challenging them to follow GL/CE).	Supporting compliance with GL/CE across all those who use computing and statistics (as per ASA and ACM) helps others comply and can limit harms/risks of harms to all.	Reputation for following CE/GL is likely to drive some collaborators away, while attracting others.	
Profession	Failures to follow GL/CE can lead to violated rights and acts against social/public good. Irreproducible results and low-quality work lower, and do not augment, informedness. Failures to follow GL/CE can undermine public trust in science, and the profession.	GL/CE represent best practices for protecting rights and promoting social/public good and informedness. Compliance with the GL/CE mitigates risks of bias and harms. Following the GL/CE identifies the trustworthy and ethical professional practitioner.	Even following GL/CE explicitly cannot avoid all possible harms. However, harms that arise when practitioners do not follow GL/CE will undermine the profession itself, and public trust in it/its work.	Commitment to GL/CE strengthens trust in the profession and enables solutions/ Conversations as well as the professional growth practitioners should support in one another.

Potential result:	HARM	BENEFIT	UNKNOWN	UNKNOWABLE
Stakeholder:				
Public/public trust	Failures to follow GL/CE can lead to violated rights and acts against social/public good. Irreproducible results and low-quality work lower, and do not augment, informedness. Public trust in science can be undermined by failures to follow professional practice standards.	Transparency and explicit respect for autonomy and prioritization of the public good supports public trust in the profession.	Transparency supports trust, even if harms do occur; following GL/CE can promote good faith.	Engagement with GL/CE promotes trust even if misuse or harms do occur.

Table 2.8.1. *Stakeholder Analysis template: Engaging in team science/teamwork*

Questions for Discussion:

Discuss similarities and differences between the ASA GLs and ACM CE principles that are discussed above, in terms of how they apply to engaging in science/work as part of a team.

What are the most striking similarities (e.g., what is unexpectedly similar)? Are there particular differences that stand out because they are A) highly predictable? B) unexpected? C) have different benefits or harms to different stakeholders?

Rank GL/CE elements from most to least important to working on a team. What does the rankability mean for you or for the discipline of statistics and data science?

Can following the GL/CE affect how you work as part of a team? How/How not?

Is the SHA in Table 2.8.1 enough to ensure ethical practice in teamwork? Why/why not?

Discuss how the ASA and ACM are, or are not, fully aligned with current policies (for your institution, organization, or lab/work team) on "research misconduct and policies for handling misconduct".

Discuss how the ASA and ACM are, or are not, fully aligned with current policies (for your institution, organization, or lab/work team) on "the scientist as a responsible member of society".

Chapter 2.9
Summary of ASA and ACM Guidance
on six tasks plus teamwork

As we saw in the preceding chapters, the statistical and the computing practitioner do not engage identically in the six tasks plus teamwork, according to their respective ethical practice standards. Specifically, we saw that "analysis" and "interpretation" are two tasks in the statistics and data science pipeline that are very different depending on whether you are (or your work is) more statistical or more computational. We also saw that there are essential dependencies of the ethical performance of all tasks on the two main/basic principles of the ASA (A. Professional Integrity and Accountability; B. Integrity of data and methods) and ACM (1. General Ethical Principles; 2. Professional Responsibilities). You will recall that these Principles also have almost complete alignment when we explored the mashup of ACM (1-2) and ASA (A-B) Principles back in Chapter 1.5.

By contrast, we also saw that the ACM CE Principle "Compliance with the Code" (Principle 4) contributed important obligations to the ethical performance of all the tasks we considered in this section (apart from Interpretation, which as we discussed, is not a specifically relevant task for the computing professional in the same way that it is for the statistical practitioner). Similarly, several aspects of the ASA GLs are relevant in every task: the overall Principles A, B, C and D, plus specific elements A2, A3, A4; B1, B2, B6, B7; D6; E4; F4, F5; G1, G5; and H2 appear to guide ethical completion of each of the six tasks plus teamwork. However, A and all its elements are essential to working on teams; B and all its elements are essential to analysis, interpretation, reporting, and working on teams; C and all its elements are essential to reporting and working on teams; D and all its elements are essential to planning/design, and working on teams. Working on teams is unique in that every single element of the ASA GLs is relevant for ethical practice of statistics and data science when working on teams.

Finally, we saw that the stakeholder analysis for each task plus teamwork/team science highlights what might be perceived as "minor" benefits for some stakeholders (you, your boss, your employer) create potentially major harms for other stakeholders (the public/public trust; the profession; unknown individuals). When we reflect back on what is essentially a massive trove of "prerequisite knowledge", we can see two things very clearly:

1. Anyone who believes that ethical quantitative practice is "simply a matter of ..." anything has not given this matter much thought at all.
2. If someone suggests to you that "all you need to do is treat other people's data as if it was your own", they also have clearly not thought about this much, because the stakeholder analysis suggests that what you do for yourself may bring minor benefits that can tend to harm others – and the profession.

By exploring the relevance of every element of the ACM and ASA ethical practice standards for each of the six tasks plus teamwork in the statistics and data science pipeline, we have shown that each practice standard is important for ethical practice along the pipeline. Some Principles and elements are always essential, while others depend on the task.

Discussion questions:

Describe how different individuals ("stakeholders") may be affected by decisions and actions. You can create a table if you like, so that all the tasks are present. Which of the six tasks plus teamwork requires the greatest attention to the stakeholder analysis, and which requires the least, and why?

Considering the six tasks plus teamwork we focused on in Section 2:
Planning/Designing
Data collection/munging/wrangling
Analysis (perform or program to perform)
Interpretation
Documenting your work
Reporting your results/communication
Engaging in team science/teamwork

- enumerate harms and benefits that are most clear to you for each stakeholder with respect to each of the 6 tasks; and
- identify which ASA GL and ACM CE Principles (and/or specific elements) seem most relevant to each task.
- Consider whether ASA and ACM have different perspectives on how to accomplish each task in the most ethical manner.

What would you say to someone who tells you that you should not spend much (or *this much*) time thinking about how to be an ethical quantitative practitioner? How do you justify your position?

Some people say that "there is no *right* answer to problems of ethical practice". Is that true? Explain your answer, taking the ACM CE and ASA GL into account.

Section 3. Ethical reasoning using ASA and ACM principles/elements: case vignettes

Chapter 3.1
Introduction to Section 3

Section 3 focuses the reader's development of Bloom's 4-6 thinking while reinforcing Bloom's 1-5 level engagement with the practice standards. In this section, there are ethical challenges that need identifying (ER KSA 3) and these come from an evaluation (Bloom's 5) of a brief vignette describing workplace events (that have actually happened). In this section the emphasis is *responding*, requiring synthesis (of your experience with your new knowledge; or of different/diverse types of knowledge) and understanding how the SHA drives the decisions as well as the justifications for how to respond to these workplace situations. The GL and CE provide guidance for planning in for each of the typical tasks in statistics and data science (Section 2) and they also support decisions and their justifications. In this section, <u>all</u> KSAs of ethical reasoning are re-introduced and utilized/practiced.

We saw that, even though our attention in Section 2 was on relatively low-ish level Bloom's attention to the GLs/CE, i.e., making sure we know that material and can think with it, none of those activities were simple! Prerequisite knowledge is quite extensive and taking a survey of what you know is very helpful in planning/avoiding ethical challenges, which forced you to build up, and then rely on, your knowledge and understanding (Bloom's 1-2) of the practice standards, in order to make plans – predicting what might happen if you behaved in ways consistent with the standards (Blooms 3-4).

However, sometimes violations of practice standards do arise, and we need to know how to respond in those cases. Together with your prerequisite knowledge, the Utilitarian and Virtue frameworks will be useful as we identify ethical issues (ER KSA 3), and then make and justify decisions (ER KSA 5) that are based on plausible alternative responses that you've considered for their consistency with the CE/GL (ER KSA 4).

Section 2 was designed to promote your engagement, and familiarity, with the Guidelines and Code of Ethics. However, as we keep repeating, just knowing what is in these guidance documents, and understanding how the guidance may differ depending on what specific task is before you, is only the "prerequisite knowledge" step of this process. Building on the earlier chapters, throughout this section, we will be utilizing all of the KSAs in ethical reasoning. For each task, we will analyze one vignette as an example, and readers are encouraged to re-do the analysis for themselves, focusing on different aspects of the practice standards. Additionally or alternatively, readers can do an analysis of that case analysis: was it helpful? What are the most likely results based on the case analysis? What do you think would be the response to such an analysis?

To prepare for the case analyses following the Ethical Reasoning KSAs, you can return to the SHA for that task from Section 2. The difference between the activities in Sections 2 and 3 is that for Section 3, there is an actual problem arising while you do a task, and so you need to apply all the KSAs to formulate a justifiable response in each case. The SHA included for each case in this section is more specific to the case, or vignette, that represents a problem arising when you set out to do the target task. You will see highlighting in the SHA that was not included when we first saw this same analysis in Section 2. Light grey highlights feature potentially substantial harms – which happen to accrue to the public/public trust; the profession; and unknown individuals. Light grey highlights draw your attention to two aspects of the harms they identify: firstly, the harms highlighted in light grey do not have the potential to be as damaging that the dark grey-highlighted harms do. Thus, the harms in the SHA are not exchangeable (as we mentioned back in Section 1). Not only are the harms not exchangeable, but also, the harms to you, your boss & employer, and your colleagues may be considered minor –particularly as compared to the other harms/potential harms. One reason for different highlighting of the harms in this way is to point out their differences/their non-interchangeability.

The second aspect of the grey-highlighted harms is even more important to our consideration: these harms – accruing to those in the workplace – can be considered to represent *incentives to ignore* ASA and ACM guidance, because ignoring the ethical practice standards can make the work go faster and with fewer complications. At worst, these behaviors "cut corners" and constitute the expedience (E4), and "statistical, scientific, and professional misconduct" that the ASA GLs warn the ethical statistical practitioner to avoid doing, condoning, and appearing to condone (H2). These behaviors are also the "poor quality work" the ACM CE warns the computing professional to resist. At best, they are simply sloppy and unprofessional. Either way, by considering how they could be seen by anyone as "legitimate" in the workplace – because they *do* speed things up – it can make the relevance of discussing the practice standards clearer to you. You may perceive that this could very well be the origins of all the detrimental research practices that exist. These are clearly not good science, not effective scientific practice, and do not yield results that contribute meaningfully to the base of knowledge. However, they do "speed things up" – i.e., there are incentives to ignore ethical practice standards. When the NIH in the US insists that only *their* topics list (Chapter 1.8) constitutes appropriate "ethical research" training, they are tending to ignore ethical practice standards – and you might wonder whether such a position encourages people who do complete NIH-approved "ethical research training", which of course does not include any mention of the ASA or ACM standards, also encourages or incentivizes NIH-trained researchers to ignore these (any, actually) ethical practice standards if they are "professional". As we saw in Chapter 1.8, the 2009 NIH policy (https://grants.nih.gov/grants/guide/notice-files/NOT-OD-10-019.html) states, "While courses related to professional ethics, ethical issues in clinical research, or research involving vertebrate animals may form a part of instruction in responsible conduct of research, *they generally are not sufficient to cover all of the above topics*." (Emphasis added) – meaning that the NIH does not deem such coursework to be sufficient or adequate. If you are an NIH-funded researcher or trainee, then you may perceive their requirement that you focus on their topics list, and their dismissal of professional ethics courses, as an *incentive to ignore* any professional ethics training (and by extension, this book).

You should also consider how these light grey-highlighted harms might be considered *disincentives to follow* ethical practice standards: If you follow the standards, then things will take longer, and they may not turn out as intended/hoped. It is a nuanced distinction, but those who don't bother to learn how professional ethical practice standards work (or that they exist/what they contain) may be/may have been incentivized to ignore them. Those who know that ethical practice standards exist, and what they include, and actively fail to follow them specifically because it is easier to do that than to follow them, are making a choice to utilize detrimental research practices (DRP) specifically because they are "better" *for them*. Such a decision obviously – and pretty dramatically, according to the NASEM 2017 report – fails the tests of justice and universality. **You should consider whether it fails the test of publication**. (That is a bit of a trick question: people who use DRP almost universally do so specifically *so they can be published*; their work, if scrutinized carefully – which is not done, in some cases, by the peer reviewers who approve such work for publication – would be seen to be incomplete and inconsistent with accepted publishing and reporting standards. So, they clearly do not have a problem *publishing* their use of DRP.)

Recall that the test of *universality* is, "would this decision work in every other similar situation?" The test of *justice* is, "does this decision affect all people equally? Would this be a fair "law" if it was enacted?" and the test of publication (originally "publicity") is, "would I like it if everyone found out that I made this decision?" (Would you want your decision published or made public in/on the news?) When you think about DRP, you should also wonder, who encourages the use or acceptability of these practices? They clearly violate ethical practice standards for statistics and data science, and computing – so **do DRP persist because too few people know about the ethical practice standards?** One reason why both the ACM and ASA standards include obligations for the ethical practitioner to make sure everyone they work with follows the standards might be to promote these ethical practices to counteract DRP.

The upshot of this discussion is: "norms" of practice in any discipline can arise from a lot of different cultural pressures – where "culture" refers to "how people in this profession 'do things'", as well as being responses to external forces. The NIH's discounting of professional ethics because they do not adequately cover their topics of interest could be considered an external force to any discipline that would cause the "norm" of ignoring ethical statistical and computing guidelines. As you have seen so far in this book, these practice standards are complex! But the point is, people might follow these norms without meaning to violate these ethical practice standards. The people who know about the ethical practice standards and ignore (or rather, violate) them because "no one will notice/know", or because it is easier and faster, benefit from incentives to ignore the practice standards. You may not know why the person in the vignettes you will see throughout Section 3 do what they do – and you might never find out. Their behavior features in these vignettes because these situations *are real*, and the ability to recognize an ethical challenge, and then formulate an ethical response, is an essential dimension to your success as an ethical practitioner.

Finally, a major difference arising from the inclusion of vignettes in this Section, as compared to the application of the ethical practice standards of the ASA and ACM to your executing of each of the statistics and data science pipeline tasks throughout Section 2, is that in Section 3, you will need to make, and justify, a **decision** – using all of the ER KSAs. This will follow our examples from Chapters 1.9 and 1.10 from Section 1 (yes, that was the last time we went through all of the KSAs of ethical reasoning!) – and also, each vignette has an analysis worked out. It is both important and instructive to consider whether the activities in Section 2, i.e., *intending* **to practice ethically,** are supportive of your decision making and the justification of the decision you do make when you are faced with any of the vignettes in Section 3 – or situations that arise as you complete tasks like these.

The final step in the ER process is to reflect on the decision you have made. Thinking about your experience reasoning your way to an ethical decision, does a decision like the one you thought through help to create the culture that promotes fluency in ethical reasoning and/or a more ethical workplace? Obviously, the steward wants the next person who has to make this or a similar decision to think it through and choose the ethical, stewardly option. Reflecting on your decision and decision-making process may increase the likelihood of the next decision maker doing so in a stewardly way

The vignettes[23] have been structured to break down what might have taken place during the Cambridge Analytica scandal, revealed on 17 March 2018 (see https://en.wikipedia.org/wiki/Facebook-Cambridge_Analytica_data_scandal for details). This scandal involved the illegal access of personal data by a company called Cambridge Analytica from Facebook users. According to reports, a data scientist at Cambridge University created a program that either did, or was modified to do, illegal harvesting of data from across Facebook (instead of just collecting what was supposed to be collected from only those who consented to have their data collected). While Cambridge Analytica was able to steal data using the program from one data scientist **who did not work there**, other statisticians and data scientists **who did work there** needed to engage in the work to collect, wrangle, munge, and then utilize the data that was collected. We know for certain that none of the professionals who were hired or paid to engage in any of these activities had been trained to be stewards of their discipline or profession, because these events all occurred prior to the date the scandal was revealed in early 2018, but stewardship is not (yet) a widely taught or practiced construct. Public trust in this discipline was harmed by each of these events, and legislation in the United States and European Union was created specifically to limit the potential for such harmful behaviors happening again. If you were faced with the decision, like those who were hired or asked to specifically modify or deploy the tool that led to illegal access of the data from millions of users, how would you have responded? The vignettes in this Section represent the use of the ER KSAs with the ACM and ASA practice standards, to provide some realistic engagement that you can reflect on. The vignettes are structured to follow the statistics and data science pipeline tasks (one per chapter):

---

[23] This paragraph reprises the argument given in Tractenberg, RE. (2019-c). *Preprint.* Becoming a steward of data science. Published in the *Open Archive of the Social Sciences (SocArXiv)*, https://doi.org/10.31235/osf.io/j7h8t.

Chapter 3.2. Planning/Designing

You are directed to design a system to scrape data from a specific source (e.g., Facebook), and are provided with specific design features of the source to ensure every data type can be scraped from every user.

Chapter 3.3. Data collection/munging/wrangling

You build a data scrape algorithm with a built-in opt-in feature, that will pop up and ask the user to opt-in to the data scraping (i.e., give consent) every time the algorithm changes, to scrape/collect more, or different data. That feature is removed.

Chapter 3.4. Analysis (perform or program to perform)

Piles of data start arriving from "the company data scraper" for you to analyze. No information is available that indicates whether consent was given for any of the data. You suspect, and then potentially identify an error in the code that removed that information.

Chapter 3.5. Interpretation

Results suggest that some people using the source (e.g., Facebook) are more susceptible to messaging (e.g., advertisements and fake news items) than others. You interpret this as signalling a need for caution/care in what your system does next – in order to limit bias, and support valid conclusions. Instead, you find reports of your work interpret the results without any caveats, and remove any suggestions that sensitivity analyses may be needed.

Chapter 3.6. Documenting your work

You are told not to document your work. When you do (because that's what the practice standards say the ethical practitioner does), your boss returns it to you with the direction, "fix this".

Chapter 3.7. Reporting your results/communication

You submit your complete and correct report of your scraping algorithm – including identification of the removal of your built in, opt-in consent to contribute data; the lack of consent accompanying data to be analyzed; and the lack of your recommendations in interpretations for limiting bias. You later discover that none of this documentation was included in the final report, but the final report is shared with stakeholders as if it is complete and correct.

Chapter 3.8. Engaging in team science/teamwork

Leadership informs your team that they bought an algorithm that you will be using to scrape data. But first, they want you to take off all the consent pop-ups, because "that ruins the user experience" and "adds personal data we will only need to strip off to preserve confidentiality".

Each of these cases walks you through the ER KSAs, and implements some of the relevant elements of the ACM and ASA practice standards that can be used to make – and justify – decisions about what to do, and how to respond, to the ethical issues that are identified. Each case also includes several questions to use in discussions with others, and to help you consider (or re-consider) whether there are other aspects of the ASA GLs or ACM CEs that might also be relevant.

Chapter 3.9. Embracing your inner ethical practitioner

Role playing: Deliver, and respond to, your case analyses in each chapter in Section 3.

In this chapter, you are encouraged to fully engage with the results of case analysis by exploring what you would do with such an analysis if you actually had to. By role-playing, you can practice initiating a conversation about an ethical challenge – utilizing your considerations at each step – with a colleague, mentor, or supervisor. In addition, you may also take the opportunity to anticipate what your colleagues, mentors, or supervisors might say in response. Will they encourage you to practice ethically? Will they try to prevent you from sharing your careful case analysis with others? How would you respond in either of these situations? You can take the opportunities in this chapter to figure out what you would do.

Chapter 3.10 Summary of Section 3 and the book: career spanning engagement in professional and ethical practice

Chapter 3.2
Planning/Designing

Case: You are directed to design a system to scrape data from a specific source (e.g., Facebook), and are provided with specific design features of the source to ensure every data type can be scraped from every user.

1. Identify and 'quantify' prerequisite knowledge:

Which ACM/ASA elements or principles seem most relevant to this vignette?

ASA:

Principle A: Professional integrity and accountability: A2
Principle D. Responsibilities to research subjects, data subjects, or those directly affected by statistical practices: D1, D4, D5, D7, D11
Principle E. Responsibilities to members of multidisciplinary teams: E1/E2
Principle F. Responsibilities to fellow statistical practitioners and the Profession: F5
Principle H. Responsibilities regarding potential misconduct: H2

ACM:

1. General Ethical Principles
2. Professional Responsibilities

Potential result: Stakeholder:	HARM	BENEFIT	UNKNOWN	UNKNOWABLE
YOU	You may appear to have intended to "steal" data if you design a system or analysis that is specifically targeted to a source like Facebook – *especially* if it appears to/actually does circumvent consent rules.	There is **no benefit** to following directions that direct you to violate professional practice standards of competence.	Your work could lead to unpredicted harms, bias, or unfair results, and although you were told to design the system/analysis incorrectly, even if you build safeguards in, *only you* will appear at fault	Inappropriately obtained data may be taken up into other applications and yield additional inappropriate and incompetent work as well as unpredictable/ unpredicted harms/bias/unfair results.
Your boss/client	Inappropriate data collection may lead to "desired" results, but they will not be reproducible or reliable, or correct. Failures to obtain consent, when discovered, will reduce faith in your boss/client.	Inappropriate data collection may lead to "desired" results (but they will not be reproducible or reliable, or correct.) Planning to collect data inappropriately is damaging.		Inappropriately obtained data may yield additional inappropriate and incompetent work as well as unpredictable/ unpredicted harms/bias/unfair results.

Potential result: Stakeholder:	HARM	BENEFIT	UNKNOWN	UNKNOWABLE
Unknown individuals	Ineffective design/ poor planning can create bias and permit privacy breaches. Results may be unfair, insecure, or both – and decisions made by incorrectly-designed systems may be unauditable. Untested – known to be incorrect – assumptions can lead to incorrect decisions and policy.			Inappropriately obtained data may yield additional inappropriate and incompetent work as well as unpredictable/unpredicted harms/bias/unfair results.
Employer	Ineffective design/ poor planning can create bias and permit privacy breaches. The company will appear to employ incompetent workers if it is discovered that inappropriate methods were used. Untested – known to be incorrect - assumptions can lead to incorrect decisions and policy.			The company will appear to employ incompetent workers- and supervisors - if it is discovered that inappropriate methods were directed to be used. A scandal may ensue when it is discovered that data has been inappropriately collected, damaging the public image of the employer.

Potential result: Stakeholder:	HARM	BENEFIT	UNKNOWN	UNKNOWABLE
Colleagues	Colleagues may assume the work was done competently, and then incorrectly build on inappropriate, biased, or otherwise unfair foundations.	Colleagues who are *not* using inappropriate methods may carry out tests that demonstrate the system cannot move forward with this design.		
Profession	The profession may appear incompetent (at best) and nefarious (at worst) when it is discovered that practitioners can or do design systems or analyses using inappropriate methodology.	There is **no benefit** to the profession that accrues from following directions that violate professional practice standards of competence.	Inappropriate work could lead to unpredicted harms, bias, or unfair results, and although you were told to design the system incorrectly, *only the practitioner* will appear at fault.	Inappropriate data may yield additional inappropriate and incompetent work as well as unpredictable/ unpredicted harms/bias/unfair results. Trust in the profession will be undermined (why not use amateurs – they're cheaper and would be expected to make the same mistakes – for less!)

Potential result:	HARM	BENEFIT	UNKNOWN	UNKNOWABLE
Stakeholder:				
Public/public trust	Inappropriate methodology can lead to conflicting results (best case) or misuse/abuse (worst case). Even if it is only one small part of a system, in inappropriate component can lead to many harms and undermine public trust in systems and the profession.			Trust in the profession will be undermined (it doesn't matter if you have a degree in that field, "those people" cannot be trusted to perform competently).

Table 3.2.1. *Stakeholder Analysis: Planning/designing with an inappropriate data scraping method*

2. Identify decision-making frameworks

From a virtue perspective, ASA Principle D clearly describes how the ethical practitioner ensures basic rights about data are respected: "The ethical statistical practitioner does not misuse or condone the misuse of data. They protect and respect the rights and interests of human and animal subjects. These responsibilities extend to those who will be directly affected by statistical practices." The ethical statistical practitioner also has responsibilities to "Use methodology and data that are valid, relevant, and appropriate..." (A2) and "Avoid condoning or appearing to condone statistical, scientific, or professional misconduct. Encourages other practitioners to avoid misconduct or the appearance of misconduct." (H2) Thus, following the directions to use a method that may inappropriately collect data is inconsistent with the ASA GLs.

ACM CE, with its utilitarian perspective, typically focuses on limiting harms but the tradeoff of harms and benefits must also be considered (as it is, implicitly, in DEFW Principle 1). The SHA shows that there are no *legitimate* benefits to following directions to use inappropriate data collection methods, and the harms are significant – to the practitioner as well as other stakeholders. While a company might benefit financially from stealing data from unknowing Facebook users, this is not a "legitimate benefit". Clearly, an entity that hides its activities in order to make a profit is unlikely to bother with SHA or professional practice guidelines; thus, the ASA Guideline element H2, "Avoids condoning or appearing to condone statistical, scientific, or professional misconduct. Encourages other practitioners to avoid misconduct or the appearance of misconduct" would be essential in this vignette to protect stakeholders from harm.

3. Identify or recognize the ethical issue

We could *assume* that the scraping method or system has "acceptable" properties, although in most cases, what is "accepted" is not generally what users or the public – data contributors or those put at risk for harm due to biased results – would call "acceptable". However, in this case, "what has always been done", i.e., using the specific scraping method or system that is known to be inappropriate, is inconsistent with the responsibilities to use best practices that are described in the GLs and CE. That is, knowingly using inappropriate methods to collect data (or do anything to/with it, really) is unethical. Using an appropriate, ethical, scraping method – or retrofitting an inappropriate system so that it *is* appropriate - is the *ethical* choice.

4. Identify and evaluate alternative actions (on the ethical issue)

The three decisions that can be made in any circumstance are: a) do nothing. b) consult or confer with a peer (ASA) or a supervisor (ACM); and c) report violations of policy, procedure, ethical guidelines, or law (ASA) or refuse to implement the system (ACM). In this case, "do nothing" means "use the given –*inappropriate*- method", and we have seen above that this is frankly unethical –so obviously, this is not a legitimate option in our analysis as it provides no real action for us to evaluate. We can change option a) to "use a different, appropriate, method or refuse to do the analysis/design the system". Option b) would become, "consult or confer with a peer (ASA) or a supervisor (ACM) as to how best to document that the recommended method is inappropriate, and use a different, appropriate, method". Option c) can be modified to, "report the use (and/or recommendation) of inappropriate methodology –which may require documentation that the recommended method is inappropriate, and use a different, appropriate, method." There have been well-publicized cases where experts using **appropriate** methodology generate results contrary to what was expected using the **inappropriate** methodology, and are harassed or otherwise punished for not violating the law and the professional practice standards (Note also that intimidation to coerce violations of practice standards are also violations of law and the GLs).

5. Make and justify a decision

Decision: *Note that all options identified in step 4 include* "use an appropriate method". The level of pressure that is brought to bear when you are told to use *inappropriate* methodology must be kept in mind as you select one of the options. Option a) is supported if you are able to make such decisions without justification or approval; so, its viability depends on your seniority or role on the team (in case you need approval to make any changes to the analysis or design). Option b) might be the *most desirable option* because consulting with a colleague/supervisor will raise awareness -beyond just you - of the fact that the specified methodology is inappropriate, but will probably also generate some documentation that this is the case. Option b) is the most desirable because it results in both a conversation about what should be done (that is *appropriate*) and some documentation about what should *not* be done, and why. This will be particularly useful if you need to convince others on the team to both utilize a different method and also engage in the opportunities for them to also innovate (changing this from a harm to a benefit according to the SHA). Option c) is an important one to consider if you encounter pressure, intimidation, or harassment from those who insist on obtaining specific data (i.e., all of it, whether consent is given or not) or results (which are guaranteed by the inappropriate method), or who insist that the inappropriate method should be used because of some expedience rather than rigor, reproducibility, or the public good. Thus, option b) may be the best choice generally, but your situation may make option c) a better choice. Certainly, the documentation of why the specified methodology is inappropriate will help others make decisions about method choice in the future.

6. Reflect on the decision

This case represents a challenge that arises – threatening to prevent you from following the ethical guidelines – from sources that are *not you*. However, clearly you must be prepared to react to challenges like these. In this case, there may be no single individual creating the ethical problem; it simply is the case that this situation -a mandate to potentially take data without consent – makes ethical practice, by you and any of your colleagues, impossible. Therefore, the decision about what to do in response to the situation must be a variant of, "use a **different, appropriate,** method". As noted, the exact option will include this, either alone (option a) or in consultation with others (option b). Option c) adds your reporting of this recommendation to use unethical means of data collection to your refusal to use that method and instead use an appropriate one.

Importantly, refusing to use that method and instead use an appropriate one passes a *test of justice* because – as the GLs and CE state – ethical work involves consideration of both applicable laws and respect for individuals to determine whether their data should be collected and used. The *test of publicity* might be embarrassing (best case) or might lead to legal action (worst case), because of the new laws that have been put into place specifically due to the Cambridge Analytica scandal, for example, and others like it worldwide. This is not a failure on the test of publicity! It only underscores how important it is to follow the practice guidelines even if you are encouraged not to at work! Finally, a *test of universality* is easily passed by all three of these alternative actions, because they all embody the GL and CE perspectives that ethical work should only-ever - use appropriate methods.

The ER KSAs, particularly KSA 4, suggest that there might be separate alternative decisions to make, and yet in this case, we could not really choose between them without more information. However, we did see that "do nothing" is not an option that is consistent with the practice standards; simply following instructions (doing nothing about the potential for data to be inappropriately taken) is unacceptable. Some form of refusing to use inappropriate technology/methods represents you taking action to ensure that you are allowed/able to follow the practice standards.

Reflection on the decision is important so that the next time someone encounters a similar situation, they will have a better idea of what to do. If you are a junior team member, option a) might be the best, or most realistic, option for you. However, you should consider what the other options mean/would require – at some point, you will be less junior, and senior enough to report behavior like this and make sure that situations like this do not occur in the future (e.g., ACM principle 3). Creating a policy that bans inappropriate data collection at your workplace (when you are more senior, or supporting it even as a junior member) would be consistent with ACM Principle 3, but is also supportive of ASA principles that articulate that the ethical statistical practitioner needs to follow the ASA GLs, and not other groups' practice standards (E1/E2; e.g., for non-members of the ASA and ACM who consider their guidelines not to apply to *them*).

Questions for Discussion:

Do you agree with the decision in this case analysis? Why or why not? Which parts of the ER process are most and least acceptable?

Discuss the relevance to this case – and your analysis/decision – of policies regarding human subjects, live vertebrate animal subjects in research, and safe laboratory practices. Are the Federal Regulations helpful (or not)? How can you incorporate what you know about these policies into any part of the case analysis (prerequisite knowledge; identifying the ethical issue; determining alternative actions; making or justifying your decision; or reflection)? Note that, if there is no alignment between such policies and this case, you can comment on that (e.g., do policies in your organization focus on *misconduct* – and ignore detrimental practices?). Do you feel that these policies could have helped in the reasoning?

Discuss the relevance to this case – and your analysis/decision – of policies in your organization relating to data acquisition and laboratory tools; management, sharing and ownership. Are the Federal Regulations relating to data acquisition, sharing and ownership in particular helpful (or not)? How can you incorporate what you know about your organizational policies or the Federal Regulations into any part of the case analysis (prerequisite knowledge; identifying the ethical issue; determining alternative actions; making or justifying your decision; or reflection)? If there are no relevant policies, you should comment on that, particularly when a) data are obtained using federal monies (grants); or b) data are collected (acquired) from humans, such that Federal Regulations about the treatment of human subjects and the acquisition of tissues and data from humans are relevant.

Keeping in mind that doing nothing/not responding are *not plausible responses,* are there other plausible alternatives (KSA 4) that you can think of? If not, discuss that; if so, list them and discuss their evaluation (are they equally consistent with GL/CE, do they lead to similar decisions, etc.).

Redo the analysis, but feature a different GL or CE principle in your reasoning process. Make sure the *justification* is different from what is given. Is the *decision* also different? Discuss your decision with its justification and the tests of universality, justice, and publicity.

Chapter 3.3
Data collection/munging/wrangling

Case: You build a data scrape algorithm with a built-in <u>opt-in</u> feature, that will pop up and ask the user to opt-in to the data scraping (i.e., give consent) every time the algorithm changes, to scrape/collect more, or different data. That feature is removed by someone.

1. Identify and 'quantify' prerequisite knowledge:

Which ACM/ASA elements or principles seem most relevant to this vignette?

ASA:

Principle A. Professional integrity and accountability: A9, A11
Principle D. Responsibilities to research subjects, data subjects, or those directly affected by statistical practices: D1, D4, D5, D7, D10, D11
Principle E. Responsibilities to members of multidisciplinary teams: E2, E4
Principle F. Responsibilities to other statisticians/statistical practitioners: F2
Principle H. Responsibilities regarding potential misconduct: H2

ACM:

1. General Ethical Principles (1.2; 1.7)
2. Professional Responsibilities (2.1; 2.3; 2.5; 2.9)

Potential result: Stakeholder:	HARM	BENEFIT	UNKNOWN	UNKNOWABLE
YOU	Using data with unknown provenance, particularly if some of it is known not to have been contributed with informed consent, violates the data contributor rights as well as ethical practice standards. The removal of your safeguard means someone on your team seeks to create – or is simply unaware of- harms to stakeholders.	While a violation of multiple specific elements of the professional practice standards and some laws, scraping data is a relatively inexpensive way to collect data.	Failing to obtain consent to use data and using stolen data (the results of breaches) could lead to unpredicted harms, bias, or unfair results, as well as risks to data contributors - with the statistician or data scientist bearing responsibility for misuse, unauthorized access, or losses of that data.	Using stolen data and data that was not contributed with informed consent may suggest to others/other system developers that collecting stolen/breached data is OK, even though this directly violates practice standards.
Your boss/client	Data with unknown provenance can lead to unpredicted harms, bias, or unfair results, and can create risks for the data contributors that were foreseeable.	Ignoring the provenance of data, while a violation of multiple specific elements of the professional practice standards, is simpler and cheaper than ensuring data are obtained with proper consent.		

Potential result: Stakeholder:	HARM	BENEFIT	UNKNOWN	UNKNOWABLE
Unknown individuals	Data that is stolen or accessed without authorization and consent (i.e., from breaches) may expose data contributors to risks, as well as to (further) misuse by others.	There are no benefits that accrue when stolen data are utilized, and no one stops the use of that kind of data.		Using stolen data and data that was not contributed with informed consent may suggest to others/other system developers that collecting stolen/breached data is OK, even though this directly violates practice standards.
Employer	Using data with unknown provenance could lead to unpredicted harms, bias, or unfair results, and can create risks for the data contributors that were foreseeable, thus incurring liability to the employer.	Ignoring the provenance of data, while a violation of multiple specific elements of the professional practice standards, is simpler and cheaper than ensuring data are obtained with proper consent.		

Potential result: Stakeholder:	HARM	BENEFIT	UNKNOWN	UNKNOWABLE
Colleagues	If others on the team are not aware that the data provenance is mixed (and some is stolen), colleagues may mistakenly share –i.e., further the misuse of- the data.	While a violation of multiple specific elements of the professional practice standards and some laws, scraping data is a relatively inexpensive way to collect it.		
Profession	The profession may appear untrustworthy when it is discovered that practitioners use whatever data they collect, even if no consent was given or even if the collected data was "available" because of a data breach. Using data with unknown provenance is a violation of multiple specific elements of the professional practice standards.	There are no benefits that accrue when stolen data are utilized, and no one stops the use of that kind of data. The individual who removed your safeguard needs to be identified and prevented from further damaging the profession (and possibly your own reputation).	Failing to obtain consent to use data and using stolen data (the results of breaches) could lead to unpredicted harms, bias, or unfair results, as well as risks to data contributors – with the statistician or data scientist bearing responsibility for misuse, unauthorized access, or losses of that data.	Using stolen data and data that was not contributed with informed consent may suggest to others/other system developers that collecting stolen/breached data is OK even though this directly violates practice standards. This decrements professional integrity in a concrete way.

Potential result:	HARM	BENEFIT	UNKNOWN	UNKNOWABLE
Stakeholder:				
Public/public trust	Public sentiment about the security of their data will continue to worsen, and people will become less inclined to contribute or sharing data if public concerns about data security, or the lack of honesty in data collection systems, continue.	There are no benefits that accrue when stolen data are utilized, and no one stops the use of that kind of data.	The fact that someone tried to prevent data from being stolen is unlikely to be shared with the public – so all practitioners will appear to be untrustworthy. Even if only one person acted to remove the safeguard, the damage to the public trust could be considerable, and ongoing.	

Table 3.3.1. *Stakeholder Analysis: Collecting data that includes un-consented and/or stolen data*

2. Identify decision-making frameworks

Data that are "scraped from the Internet" cannot be considered to have been contributed to the planned analyses or system with consent. However, some data (e.g., counts of persons walking through a turnstile) simply cannot be deemed "data you must consent to contribute". The problem in this case is that the data did not come from sources like this – and you *had included a provision for consent* – presumably because as an ethical practitioner, you were following D1: "Keeps informed about and adheres to applicable rules, approvals, and guidelines for the protection and welfare of human and animal subjects. Knows when work requires ethical review and oversight" - but that consent mechanism was removed.

As Principle D states, "The ethical statistical practitioner does not misuse or condone the misuse of data. They protect and respect the rights and interests of human and animal subjects. These responsibilities extend to those who will be directly affected by statistical practices." Specifically, the ethical statistical practitioner has responsibilities to "Uses data only as permitted by data subjects' consent when applicable or considering their interests and welfare when consent is not required. This includes primary and secondary uses, use of repurposed data, sharing data, and linking data with additional data sets." (D5). Moreover, the ethical statistical practitioner "Does not conduct statistical practice that could reasonably be interpreted by subjects as sanctioning a violation of their rights. Seeks to use statistical practices to promote the just and impartial treatment of all individuals." (D11). Thus, utilizing data with unknown provenance violates the ASA GLs -meaning anyone who uses the data that was collected once your consent feature was removed will violate the ASA GLs. Obviously, failing to protect basic human rights by using data you do not have consent to use violates both D5 and D11 (among others), but also violates ASA GL Principle H2 ("Avoids condoning or appearing to condone statistical, scientific, or professional misconduct. Encourages other practitioners to avoid misconduct or the appearance of misconduct."). The individual who removed the consent feature is engaging in misconduct of all three types – and is creating the situation where others may also engage in that misconduct if they use the data. Thus, pointing out this created problem will "help strengthen the work of others through appropriate peer review" (F2). The virtue perspective has a lot of guidance to offer in this case.

The ACM CE, with its utilitarian perspective, focuses on limiting harms and in this case, there are many potential harms, and no legitimate benefits, to using data that was contributed to your analysis without consent, or using data with unknown provenance (contrary to ASA D10). ACM CE supports correcting –as well as overcoming – the removal of your opt-in feature. This is the only/best way to avoid real harms and risks of harms that will accrue to both the individuals whose data is scraped in violation of their rights to privacy and confidentiality –harming *them*, and also *harms to those rights*. When organizations act as if it is OK to use data they obtain without consent, it strengthens a "norm" to do so – harming rights against this; computing professionals have a responsibility to ensure that their organization strengthens the ACM CE as norms (Principle 3), and that methods that end up violating human rights do not become norms (Principle 4).

3. Identify or recognize the ethical issue

Both the ASA and ACM standards recognize the basic human right to determine how one's data is used, and to be informed about that prior to the use of that data, so both standards lead to the recognition that, without the consent feature you originally built into the scraping system, the data that is scraped will have been unethically sourced. This is, effectively, a waste of the resources used to collect the data – because it won't be useable by any ethical practitioner. Both the ASA and ACM practice standards state explicitly that professionals must respect laws and practice standards to ensure data are collected with the knowledge and consent of contributors. The primary ethical issue in this case is that both GLs and CE require that data be obtained with consent and the contributors' knowledge of what the data will be used for, but someone removed this from the system without the designer's knowledge. A secondary issue is that the person who removed the consent feature needs to be told that that is inappropriate and leads to multiple violations of law and the practice standards, as well as the respect for autonomy that is an essential human right worldwide.

4. Identify and evaluate alternative actions (on the ethical issue)

The three decisions that can be made in any circumstance are, again: a) do nothing. b) consult or confer with a peer (ASA) or a supervisor (ACM); and c) report violations of policy, procedure, ethical guidelines, or law (ASA) or refuse to implement the system (ACM). Clearly, "do nothing" is both totally inappropriate in this case, and also a violation of the practice standards relating to respecting data contributors (as well as violating respect for autonomy if your organization uses the system without the consent feature). There is also the fact that someone took your correct/appropriate methodology and changed it so that it is now not appropriate – putting the entire organization at risk. The ACM CE specifies that simply not participating or withdrawing once the unavailability of provenance of the data becomes known is not acceptable/not sufficient. We can change option a) from "do nothing" to "stop the progress of this project, and stop any use of data resulting from its use, until the consent feature can be re-implemented". Then option b) would become, "consult or confer with a peer (ASA) or a supervisor (ACM) as to how best to stop the progress of this project, and stop any use of data resulting from its use, until the consent feature can be re-implemented". Option c) must also be modified, because simply reporting the fact that the data collection violates the GLs (ASA), or refusing to implement the system (that scrapes up stolen/breached data) (ACM) might not ensure that the data with unknown provenance is both not used and also, its collection ceases. So, option c) needs to change to "report violations of policy, procedure, ethical guidelines, or law to a peer (ASA) or a supervisor (ACM), and stop the progress of this project, and stop any use of data resulting from its use, until the consent feature can be re-implemented."

5. Make and justify a decision

Decision: *Note that all options include* "stop the progress of this project, and stop any use of data resulting from its use, until the consent feature can be re-implemented". That is because the core of response to this situation is to **follow the practice standards** and to ensure that your organization will only utilize data that has the consent of the contributors. What is less clear is what *else* to do. Option a) says "follow the ethical practice standards, and do not use inappropriately-obtained data". This would be the *least desirable option* because the data collection – as well as its use – both need to stop, because both are inappropriate. However, if data with unknown provenance is commonly collected in your work context, and that is why your consent feature was removed (no matter whether it is illegal, unethical, or both) it might be impossible for you to do anything ethical except seek another team, other colleagues, or a different job. Option b) would be the *most desirable option* in that consulting with a colleague/supervisor will ensure at least some kind of notification that, or publicity for the fact that, data collection without consent is frankly unethical, as is the use of such inappropriately obtained data. There may not be policies in your workplace against such interference/violation of the ASA GLs or ACM CE, so reporting (option c) may not be a viable option, making it a slightly less desirable option than b). However, in contexts where such policies do exist, it may still be difficult to report such behaviors; but you (the practitioner) can – and should - still demonstrate your professional competence and follow the relevant practice standards. Option c may not be as much about "reporting" a person or practice as it is about "educating" people or your company/organization about the practice standards – particularly, ensuring that people do recognize that you included a consent feature because it is both ethical and consistent with consent rules or laws (depending on the context). It is important *to the profession* to support the next practitioner who ends up in a similar situation (or better, to prevent such unethical data collection in the future), so it is definitely worth considering politely notifying relevant individuals that the data collection system violates (or violated, before you stopped it) the ACM CE, and you cannot be directed to violate the CE and/or the GLs. When you present the system and it features modifications that may limit the amount of data but also limit the likelihood of including stolen/breached data, it would be an opportunity to point out to everyone involved, and all stakeholders, that this is the only way to meet ASA GLs/ACM CE, respect fundamental human rights, and limit harms/risks of harms. Note also that limiting harms/risks of harms in these ways complies with ASA A9 ("Takes appropriate action when aware of deviations from these Guidelines by others").

6. Reflect on the decision

The decision to choose option b or c would be based on the fact that for both computing and statistics, the ethical practitioner protects human rights – and this is true whether or not harms or risks of harms are likely (or especially likely) to accrue. In this case, the ethical practitioner must be especially concerned about the tests of "universality" and "justice", because confidentiality and privacy breaches are becoming common, but the use of data from those breaches should _not_ become more common. When organizations act as if it is OK to use data they _obtain without consent_, it strengthens a "norm" to do so – harming human rights by both violating them and also by reducing the seriousness of such violations. While computing professionals, but not statisticians, have a responsibility to ensure that their organization strengthens the ACM CE as norms (Principle 3), and that methods that end up violating human rights do not become norms (Principle 4), statisticians do have the responsibility to protect human rights (ASA D). The tests of universality and justice are important features of this decision, even though these are not specifically part of either practice standard.

Note that there _are_ cases where data might be collected without any need for/mechanism for consent; one example might be being-counted-by-a-turnstile; or cameras on roads in the US that capture people exceeding the speed limit. Driving down a public road is a public act, and while people might not want it publicized that they got a speeding ticket, individuals can have no claim or expectation to privacy when breaking the speed law in such a public way. In this case, the individual (you) designed the system with an opt-in consent feature – suggesting that there was a reasonable expectation of such a feature. No ethical practitioner would then analyze or even consider working with data with such provenance; this is the specific subject of ASA GLs D5 and

D11. ACM CE 1.6 also addresses the considerations of provenance that computing professionals should give. Thus, whomever removed the consent feature has violated these guidelines and also acted in a way that is inconsistent with both professional practice standards. It is not clear why they removed your consent feature, but that needs to be considered, and addressed, although it is not the primary ethical challenge in this vignette. Also, options b) and c), both of which are preferable to a), involve some kind of reporting or notification or discussion – these are preferred options specifically because they do stop the collection and use of data without consent, but they also address the fact that the consent feature was removed. Without addressing that fact, this could happen again to someone with less awareness of their professional obligations, so your initiating discussions about the importance of respecting the fundamental human right of autonomy can be a critical contributor to changing the culture at your organization and informing your colleagues of this responsibility.

This case is important because of the fact that in some situations, the exact same case may lead to an ethical dilemma (where the only options are "prevent the data from being collected" or "ensure that data contributors are aware that their data are not secured, and hope they consent anyway") or to *no ethical dilemma* at all (in the case where there can be no reasonable expectation of privacy or security of their data, such as the turnstile count). This is one example of what both the GLs and CE note in their preambles: there is no specific rule or algorithm that can apply in every case, and the ethical practitioner is familiar with the ethical practice standards and also knows how and when to apply them.

In some contexts, data are collected without the knowledge of the data contributor, but this is contrary to both the ASA GLs and the ACM CE: both state that human data contributors must consent to the collection and use of their data (although the implication is that consent is required whenever it can be reasonably expected). There are laws and regulations about informing data contributors that their data are being collected, and ensuring they consent to that data contribution. Since 2018, many websites that collect data from site visitors require that you consent or else you will not be able to visit the site. There may also be institutional policies –particularly in academia and the pharmaceutical industry – relating to considerations of the rights of all humans to determine what data is collected from them; these may be specifically important for biomedical work and for the types of research involving human subjects that are common in academic work. Even if a group, organization, or company "usually" collects data without informed consent from the contributors (e.g., Amazon collecting data on purchases), it does not relieve them of the burden of *securing* that data if the contributor has a reasonable expectation that the data should be secured. There are fundamental responsibilities that any and all data scientists and statisticians bear – as well as those who use computing and statistics, per the CE and GLs – to respect these fundamental human rights.

Questions for Discussion:

Do you agree with the decision in this case analysis? Why or why not? Which parts of the ER process are most and least acceptable?

Discuss the relevance to this case – and your analysis/decision – of policies regarding human subjects, live vertebrate animal subjects in research, and safe laboratory practices. Are the Federal Regulations helpful (or not)? How can you incorporate what you know about these policies into any part of the case analysis (prerequisite knowledge; identifying the ethical issue; determining alternative actions; making or justifying your decision; or reflection)? Note that, if there is no alignment between such policies and this case, you can comment on that (e.g., do policies in your organization focus on *misconduct* – and ignore detrimental practices?). Do you feel that these policies could have helped in the reasoning?

Discuss the relevance to this case – and your analysis/decision – of policies in your organization relating to data acquisition and laboratory tools; management, sharing and ownership. Are the Federal Regulations relating to data acquisition, sharing and ownership in particular helpful (or not)? How can you incorporate what you know about your organizational policies or the Federal Regulations into any part of the case analysis (prerequisite knowledge; identifying the ethical issue; determining alternative actions; making or justifying your decision; or reflection)? If there are no relevant policies, you should comment on that, particularly when a) data are obtained using federal monies (grants); or b) data are collected (acquired) from humans, such that Federal Regulations about the treatment of human subjects and the acquisition of tissues and data from humans are relevant.

Keeping in mind that doing nothing/not responding are *not plausible responses,* are there other plausible alternatives (KSA 4) that you can think of? If not, discuss that; if so, list them and discuss their evaluation (are they equally consistent with GL/CE, do they lead to similar decisions, etc.).

Redo the analysis, but feature a different GL or CE principle in your reasoning process. Make sure the *justification* is different from what is given. Is the *decision* also different? Discuss your decision with its justification and the tests of universality, justice, and publicity.

Chapter 3.4
Analysis (perform or program to perform)

Case: Piles of data start arriving from "the company data scraper" for you to analyze. You begin to run existing analysis programs, but discover that no information is available to indicate whether consent was given for any of the data. You suspect, and then potentially identify, an error in the analysis code that removed that information.

1. Identify and 'quantify' prerequisite knowledge:

Which ASA/ACM elements or Principles seem most relevant to this vignette?

RECALL: NB: "analysis" has different meanings for straight-computation (ACM) and straight-statistical (ASA) users, with blended meanings for those who use computation extensively in their statistical practice (or who use statistics in their computing practice), i.e., data scientists. Specifically, ASA GLs refer to the analysis of data, and in some cases to an analysis of how people use the data or the analysis (evaluation) of others' use of data, results, or methods. By contrast, the ACM CE refers to the analysis of risks, the evaluation (and implied analysis) of systems or their plans/designs, and to the analysis of activities, or the effects of computing on stakeholders.

Despite the different perspectives of the ASA and ACM on "analysis", **the majority of Principles in each practice standard are relevant in this case:**

ASA:

Principle A. Professional Integrity and Accountability: A3, A9, A11
Principle B. Integrity of data and methods: B1, B6
Principle D. Responsibilities to research subjects, data subjects, or those directly affected by statistical practices: D; D1, D4, D5, D10, D11
Principle E. Responsibilities to members of multidisciplinary teams: E1, E2, E4
Principle F. Responsibilities to fellow statistical practitioners and the Profession: F5
Principle H. Responsibilities regarding potential misconduct: H2

ACM:

1. General Ethical Principles (1.1)
2. Professional Responsibilities (2.1; 2.4; 2.5; 2.9)
3. Professional Leadership Responsibilities (3.5)

Potential result: Stakeholder:	HARM	BENEFIT	UNKNOWN	UNKNOWABLE
YOU	Confirming, documenting, and then fixing the error will take time, adds effort; may identify other problems that require additional time/effort. Failing to fix the error may save time but lead to unpredictable propagation of that error throughout the workflow.	Correction of the error means future work with the code will be facilitated –and correct, and interpretable. Identifying the error may limit the collection and use of inappropriately- or unethically sourced data in analyses (by you and others).	Competent analysis (ASA) supports reproducible results; effective analysis (ACM) promotes public good.	
Your boss/client	Identifying, documenting, and then fixing the error will take time, adds effort; may identify other problems that require additional time/effort. This error may be discovered in a 'product' that the client has paid for in the past, identifying past errors and undermining this and other projects	Correction of the error means future work with the code will be facilitated –and correct, and interpretable. The correction of this error may limit the future collection and use of inappropriately- or unethically sourced data by the organization.	Identification of error may require resources to fix/address. *Appropriate analysis* –once the error is fixed - may not yield desired results.	

Potential result: Stakeholder:	HARM	BENEFIT	UNKNOWN	UNKNOWABLE
Unknown individuals	Analysis required or ACM CE violated; weak statistical analysis undermines reproducibility and rigor.	Identification and correction of the error allows mitigation of risks of harms (ACM); supports valid inferences (ASA), and can limit the unpredictable propagation of the error to other parts of the workflow.		
Employer	Identifying, documenting, and then fixing the error will take time, adds effort; may identify other problems that require additional time/effort. This error may be discovered in a 'product' that the client has paid for in the past, identifying past errors and undermining this and other projects	Correction of the error means future work with the code will be facilitated –and correct, and interpretable.	Identification of error may require resources to fix/address. *Appropriate* analysis –once the error is fixed - may not yield desired results.	Identification of the error may highlight past payments or purchases/sales of incorrect/error-prone work. Liability may be incurred.

Potential result: Stakeholder:	HARM	BENEFIT	UNKNOWN	UNKNOWABLE
Colleagues	The discovered error may complicate colleagues' work.	Identification and correction of the error allows mitigation of risks of harms (ACM); supports valid inferences (ASA), and can limit the unpredictable propagation of the error to other parts of the workflow.		Project could be delayed until risks of additional errors, and resulting harms/ potential harms are addressed.
Profession	Identifying, documenting, and then fixing the error will take time, adds effort; may identify other problems that require additional time/effort. Failing to fix the error may save time but lead to unpredictable propagation of that error throughout the workflow.	Correction of the error means future work with the code will be facilitated –and correct, and interpretable. This supports ethical and competent practice by future practitioners.	Competent analysis (ASA) supports reproducible results; effective analysis (ACM) promotes public good.	Commitment to identification and correction of the error, consistent with GL/CE, might strengthen trust in profession. The error itself can enable innovation (to fix, avoid, or detect similar errors in future), as well as conversations about ethical practice.

Potential result:	HARM	BENEFIT	UNKNOWN	UNKNOWABLE
Stakeholder:				
Public/public trust	Even if errors are noted so they can be fixed, weak or error-prone analysis, or analyses based on unethically sourced data, undermines public trust in reproducibility and rigor, as well as the profession.	Identification and correction of the error allows mitigation of risks of harms (ACM); supports valid inferences (ASA) and can limit the unpredictable propagation of the error to other parts of the workflow and decisions based on them.	Transparency supports the public trust, even if risk/harms are identified; can tend to inform the public about limitations inherent in the profession.	

Table 3.4.1. *Stakeholder Analysis: you discover an error in an analysis program you are running*

2. Identify decision-making frameworks

As noted, ASA GLs represent a virtue approach, which supports correcting the program so that the analysis is based solely on ethically sourced data, so that it "Protects people's privacy and the confidentiality of data concerning them, whether obtained from the individuals directly, other persons, or existing records." Because the ethical practitioner "Knows and adheres to applicable rules, consents, and guidelines to protect private information." (D4). Moreover, D5 specifies that the ethical practitioner "Uses data only as permitted by data subjects' consent when applicable or considering their interests and welfare when consent is not required. This includes primary and secondary uses, use of repurposed data, sharing data, and linking data with additional data sets." (D5). Consent -or the lack of it - aside, the ethical statistical practitioner also "Avoids compromising scientific validity for expediency" (E4), so although it would be more expedient to let the system run on with the known error, E4 and Principle B dictates that the ethical statistical practitioner: "Strives to promptly correct any errors discovered while producing the final report or after publication. As appropriate, disseminates the correction publicly or to others relying on the results." (B5). This error exists in the code that was given to you, and is not of your making. However, this particular error makes it impossible to fulfill multiple other ASA principles, specifically those relating to respecting the basic human right to give consent for data collection and ensuring that the provenance of data is known before the ethical statistical practitioner can use that data (D1, D4, D5, D11). The virtue perspective brings multiple aspects of ethical practice to bear on this ethical challenge.

ACM CE takes the utilitarian perspective, reiterating throughout the practice standards that the computing professional has a responsibility to analyze the work they do/are asked to do so that decisions made on the basis of the system or its subparts are *justifiable*. If an error is identified in the analysis part of the system, then results may be biased or incorrect, potentially causing predictable (once the error is identified) but fixable harms. More specifically, "1.3 A computing professional should be transparent and provide full disclosure of all pertinent system capabilities, limitations, and potential problems to the appropriate parties." This code was given to you, and you have identified a potential problem with that code; as long as that code has been being used, it has been creating a problem. Thus, the utilitarian perspective highlights the importance of correcting the error for yourself and the wider community.

3. Identify or recognize the ethical issue

Both the ASA and ACM standards support the fact that the discovery of an error in the code – particularly one that might lead to unethically-sourced data being analyzed (ASA), is a significant ethical issue. As an expert in your field, you identified the error. The GLs and CE specify that identifying it is not sufficient – an additional ethical issue is raised if you do not fix the error. The ASA GLs stipulate that corrections should be disseminated, and the ACM CE specify similarly that limitations of the system should be disclosed (unless the error is fixed, in which it is no longer a system limitation). Since the code, or the data the code collects, may have been used in (many) prior analyses, there could be longstanding problems that the error you discovered has created. Thus, even if the error in the code is fixed and there is no longer a system limitation, the prior work with that system may have been compromised; so, the ACM CE imply that the discovery of the error and its correction both need to be disclosed; ACM 1.2 states, "To minimize the possibility of indirectly or unintentionally harming others, computing professionals should follow generally accepted best practices unless there is a compelling ethical reason to do otherwise." It is essential that the "accepted best practices" are literally "best", and not simply what your organization typically does – particularly if you work somewhere where consent is routinely ignored; this is clearly *not the best* (nor is it legal) practice, even if it is "accepted" at your workplace. The ASA GLs are explicit that a new ethical issue is raised if the error is not disclosed, while the ACM CE is not explicit about this, but it can be inferred.

4. Identify and evaluate alternative actions (on the ethical issue)

The three decisions that can be made in any circumstance are, again: a) do nothing. b) consult or confer with a peer (ASA) or a supervisor (ACM); and c) report violations of policy, procedure, ethical guidelines, or law (ASA) or refuse to implement the system (ACM). Clearly, "do nothing" is both totally inappropriate in this case, and also a violation of the practice standards relating to their outlined responsibilities to show respect for data contributors (as well as violating respect for autonomy if your organization uses the system without the consent feature. We can change option a) from "do nothing" to "stop the progress of this project, and stop any use/analysis of data resulting from its use, until the consent indicator error is fixed". Then option b) would become, "consult or confer with a peer (ASA) or a supervisor (ACM) as to how best to stop the progress of this project, and stop any use/analysis of data resulting from its use, until the consent indicator error is fixed". Option c) must also be modified, because simply reporting the fact that the data collection violates the GLs (ASA), or refusing to implement the system (that scrapes up stolen/breached data) (ACM) might not ensure that the data with unknown provenance is both not used and also, its collection ceases. So, option c) needs to change to "report violations of policy, procedure, ethical guidelines, or law to a peer (ASA) or a supervisor (ACM), and stop the progress of this project, and stop any use/analysis of data resulting from its use, until the consent indicator error is fixed." Consulting a peer or supervisor is an essential feature of all options if the decision to correct the program requires approval, i.e., if you are junior. However, the notification aspect of the solution may very well require approval and/or supervisory consultation, if the error you discovered has been in place long enough to have featured in actual decisions already having been made. The ACM CE is explicit about your consideration of the aftereffects of your reporting an error. In 1.2, the CE states, "A computing professional has an additional obligation to report any signs of system risks that might result in harm. If leaders do not act to curtail or mitigate such risks, it may be necessary to "blow the whistle" to reduce potential harm. However, capricious or misguided reporting of risks can itself be harmful. Before reporting risks, a computing professional should carefully assess relevant aspects of the situation." In this case, unethically sourced data may be collected and analyzed using this code. You have identified a specific bit of code that makes it impossible for anyone to know if consent was obtained properly, and that is a literal "system risk that might result in harm" – as well as representing specific disregard of data contributor rights, and creating a situation where others may unknowingly use data that is unethically sourced. This code error creates harms to data contributors and data users, so notification ("blowing the whistle") is a clear responsibility according to the ACM CE.

5. Make and justify a decision

Decision: *option b* ("consult or confer with a peer (ASA) or a supervisor (ACM) as to how best to stop the progress of this project, and stop any use/analysis of data resulting from its use, until the consent indicator error is fixed".) The error must be corrected, and the error and its resulting impact in creating potentially unethically sourced data should be reported throughout the organization, as supported by both ASA and ACM standards. This is the heart of professional integrity and accountability (ASA Principle A) and general ethical principles (ACM Principle 1). It is also supportive of ethical practice by you and others (mainly ASA Principles A, B, D), and ACM Principle 2 ("Computing professionals should insist on and support high quality work from themselves and from colleagues.").

6. Reflect on the decision

The most important part of this decision-making procedure is that "do nothing" is absolutely unacceptable. Professional integrity and accountability (ASA Principle A) demand that errors in our work be identified and corrected, while ASA F2 recommends strengthening others' work with constructive feedback. If results have already been published, it may be possible to publish an erratum – an amendment to the original publication. If an error evolves because technology or information improves, for example, more sensitive detection methods become available; or it is determined that subgroups behave/respond differently, then the correction to the original is even more important. In the identification of alternative actions, we noted that stopping the collection and analysis of potentially unethically sourced data is the core of all three options. While the ACM CE clearly specify that leadership should be notified whenever that is appropriate, in this case you are determining what to do about an error that existed before you were handed the code. Thus, the decision to correct the program may require approval, because it is possible that results from analyses of the "corrected" code will differ substantively from the original; corrections to the code you are working with may also need to be propagated to other code as well. The notification aspect of the solution may very well require approval or supervisory consultation, if the error you discovered has been in place long enough to have featured in actual decisions already having been made (which was true for the original case!).

The SHA shows plainly that, because all options include stopping the use of the code with the error, and correcting it, all of these options pass a *test of justice* (and, not surprisingly, doing nothing clearly fails the test of justice). The SHA suggests that the harms that can accrue to the profession and the public trust – as well as to you - require careful consideration of the *test of publicity*: you would not want it publicized that you either purposefully steal data, nor that you simply follow directions without knowing (or apparently caring) whether or not the directions are appropriate for the problem and the data. Neither of these would be pleasant if shared publicly, but there may be legal ramifications if it became known that you/your organization set out to purposefully steal data and circumvent consent and basic human rights laws; thus, because publicizing the organization's purposeful circumvention of laws is clearly a far less desirable public result than making it public that you detected an error in the code and fixed it, all options pass the test of publicity (and doing nothing clearly fails it). The SHA shows the substantial harms (with minimal benefits) that accrue in this case if you did not take any of the options; doing nothing clearly fails the *test of universality*. However, the notification aspects of options b) and c) tend to make the universality of these options "better" – because not only would you put an end to the collection and analysis of unethically sourced data (i.e., protecting just those contributing data), you would also be alerting your colleagues and the wider community that the error existed (and so, data collected prior to the fix should not be used) in option b), and possibly changing the policy of the organization (in option c) to ensure that such errors were detected and fixed more regularly in future. Thus, all three options pass a test of universality, but b) and c) broaden the "universe" of those who benefit.

Obviously, all professionals strive to do their best work, and so, discovering an error can be embarrassing. Importantly, this case might have included a previously undiscovered error, which would not represent malicious "vandalism" by the coder, nor a bad-faith actor who has been disincentivized from following the ethical practice standards. Reflecting on that, and taking into consideration the real harms that can accrue when honest errors in analysis are uncorrected, it might be useful to implement a peer-review system in your team or organization; this is consistent with both ASA and ACM practice standards while also possibly encouraging more, and more open, discussion of how the systems or analyses are planned and executed. However, in the introduction to this section, we stated that the cases all have to do with the tasks associated with the code that was used to literally steal data from unconsenting Facebook users. Thus, this undiscovered "error" may actually have been inserted very purposefully - maliciously. It may have been unknown to all the data scientists at the company that someone had intended to scrape all data, without consideration of consent, from Facebook, and purposefully inserted this error. Whether or not the error in the code was placed there inadvertently or maliciously, code review may be an important feature to suggest or implement in your organization or team. The work of famous statistician William Edwards Demming encourages all members of a team to take responsibility for the end result of any system, such that if any one of them identified a problem or error, they could – and would be encouraged to – stop the system to figure out how to correct the error (https://en.wikipedia .org/wiki/W._Edwards_Deming). This is referred to as "quality control", and represents a desirable "norm" for quantitative practice, because it encourages quality improvement, and charges all members of a team equally with promoting this improvement (without recognizing errors, they cannot be fixed or addressed). So, while the "disseminating" (ASA) or "disclosing" (ACM) aspect of fixing the error in this vignette might seem daunting and possibly embarrassing, describing that effort (and doing it!) as a "quality improvement" step may serve to incentivize such behavior – simultaneously promoting compliance with ASA and ACM practice standards.

Questions for Discussion:

Do you agree with the decision in this case analysis? Why or why not? Which parts of the ER process are most and least acceptable?

Discuss the relevance to this case – and your analysis/decision – of policies regarding conflicts of interest (COI) in your organization. The COI may be for you, but may also be for others on your team (reporting an error may seem contrary to people's careers!). How can you incorporate what you know about these policies into any part of the case analysis (prerequisite knowledge; identifying the ethical issue; determining alternative actions; making or justifying your decision; or reflection)? Note that, if there is no alignment between such policies and this case, you should comment on that (e.g., do policies in your organization focus on *avoiding* COI or *managing* them?). Do you feel that policies or thinking about COI could have helped in the reasoning of this case?

Discuss whether this case represents an opportunity for those involved to reflect on "the scientist or practitioner as a responsible member of society". If so, is it limited to "society" defined as the community of practitioners in the same field (i.e., to "the profession"), or is the case also (or only) relevant to the scientist or practitioner as a responsible member of the wider society?

Keeping in mind that doing nothing/not responding are *not plausible responses*, are there other plausible alternatives (KSA 4) that you can think of? If not, discuss that; if so, list them and discuss their evaluation (are they equally consistent with GL/CE, do they lead to similar decisions, etc.).

Redo the analysis, but feature a different GL or CE principle in your reasoning process. Make sure the *justification* is different from what is given. Is the *decision* also different? Discuss your decision with its justification and the tests of universality, justice, and publicity.

Chapter 3.5
Interpretation

Case: Results suggest that some people who are the source of data your organization wants to use (e.g., Facebook) are more susceptible to messaging (e.g., advertisements and fake news items) than others. You interpret this as signalling a need for caution/care in what your system does next with this data, and you include this in all your communication – in order to limit bias, ensure that no stakeholders are misled, and support valid conclusions resulting from your statistical, computing, and data science practice. Instead, you find reports of your work interpret the results without any caveats, and remove any suggestions that caution or sensitivity analyses may be needed.

1. Identify and 'quantify' prerequisite knowledge:

Which ASA/ACM Principles (and/or specific elements) seem most relevant to this vignette?

NOTE: Like "analysis", "interpretation" has different meanings for straight-computation (ACM) and straight-statistical (ASA) users. The meaning for statisticians and data scientists in terms of the ASA GLs are clearly referring to interpreting the analysis of data or in the case of meta-analysis, the interpretation of a collection of evidence (and data). The meaning of "interpretation" in the statistical sense will involve competent assessment of data, methods, and results. By contrast, because the ACM CE uses the term "analysis" (or related ideas) to refer to the analysis of risks, the evaluation (and implied analysis) of systems or their plans/designs, to the analysis of activities, or to the effects of computing on stakeholders, we consider that the coverage of the ACM CE on "analysis" sufficiently addresses the ACM's perspective on how the computing professional should utilize (*interpret*) the results of those analyses. This analysis, then, focuses only on utilizing the ASA GLs and their specific perspectives on interpretation, relating more particularly to the interpretation of statistical analyses. As the interpretation of statistical analysis is relevant in any computing system, the ASA GLs should be sufficiently informative for the ACM member/individual who utilizes the ACM CE to guide their decision-making.

Despite the different meaning of "interpretation" for users of the ASA and ACM standards, <u>all</u> Principles from the ASA perspective and one from ACM are relevant in this case:

ASA:

Principle A: Professional Integrity and Accountability: A4, A9

Principle B. Integrity of data and methods: B2, B3, B6

Principle C. Responsibilities to stakeholders: C2, C4

Principle D. Responsibilities to research subjects, data subjects, or those directly affected by statistical practices: D; D6, D10

Principle E. Responsibilities to members of multidisciplinary teams: E2, E4

Principle F. Responsibilities to fellow statistical practitioners and the Profession: F4

Principle G. Responsibilities of leaders, supervisors, and mentors in statistical practice: G1, G5

Principle H. Responsibilities regarding potential misconduct: H2

ACM:

2. Professional Responsibilities

Potential result: Stakeholder:	HARM	BENEFIT	UNKNOWN	UNKNOWABLE
YOU	Identifying, documenting, and then fixing the errors of interpretation will take time, adds effort; may identify other problems that require additional time/effort. Failing to fix the errors of interpretation may save time but lead to invalid results, unsupported next steps, and unpredictable propagation of errors throughout the workflow.	Correction of the errors of interpretation means future work with the code will be facilitated – and correct, and interpretable.	Competent interpretation (ASA) supports reproducible results; effective analysis (ACM) promotes public good.	

Potential result: Stakeholder:	HARM	BENEFIT	UNKNOWN	UNKNOWABLE
Your boss/client	Identifying, documenting, and then fixing the errors of interpretation will take time, adds effort; may identify other problems that require additional time/effort. The errors of interpretation may be discovered in a 'product' that the client has paid for in the past, identifying past errors and undermining this and other projects	Correction of the errors of interpretation means future work with the code will be facilitated —and correct, and interpretable.	Identification of errors of interpretation may require resources to fix/address. *Appropriate* interpretation may not yield desired results.	
Unknown individuals	Analysis required or ACM CE violated; inappropriate interpretation undermines reproducibility and rigor, and limits generalizability.	Identification and correction of the errors of interpretation allows mitigation of risks of harms (ACM); supports valid inferences (ASA), and can limit the unpredictable propagation of the error to other parts of the workflow.		

Potential result: Stakeholder:	HARM	BENEFIT	UNKNOWN	UNKNOWABLE
Employer	Identifying, documenting, and then fixing the errors of interpretation will take time, adds effort; may identify other problems that require additional time/effort. The errors may be discovered in a 'product' that the client has paid for in the past, identifying past errors and undermining this and other projects	Correction of the errors of interpretation means future work will be facilitated –and correct, and interpretable.	Identification of errors of interpretation may require resources to fix/address. *Appropriate* interpretation may not yield desired results.	Identification of errors of interpretation may highlight past payments or purchases/sales of incorrect/error-prone work. Liability may be incurred.
Colleagues	The discovered errors of interpretation may complicate colleagues' work.	Identification and correction of the errors of interpretation allows mitigation of risks of harms (ACM); supports valid inferences (ASA), and can limit the unpredictable propagation of the errors to other parts of the workflow.		Project could be delayed until risks of additional errors, and resulting harms/ potential harms are addressed.

Potential result: Stakeholder:	HARM	BENEFIT	UNKNOWN	UNKNOWABLE
Profession	Identifying, documenting, and then fixing the errors will take time, adds effort; may identify other problems that require additional time/effort. Failing to fix the errors may save time but lead to unpredictable propagation of that errors throughout the workflow and into the community.	Correction of the errors of interpretation means future work with the code will be facilitated –and correct, and interpretable. This supports ethical and competent practice by future practitioners.	Competent interpretation (ASA) supports reproducible results; effective analysis (ACM) promotes public good.	Commitment to identification and correction of the errors of interpretation, consistent with GL/CE, might strengthen trust in profession. The errors themselves can stimulate innovation (to fix, avoid, or detect similar errors in future), as well as conversations about ethical practice.
Public/public trust	Even if errors of interpretation are noted so they can be fixed, weak or error-prone analysis undermines public trust in reproducibility and rigor.	Identification and correction of the errors of interpretation allows mitigation of risks of harms (ACM); supports valid inferences (ASA), and can limit the unpredictable propagation of the error to other parts of the workflow and decisions based on them.	Transparency supports the public trust, even if risk/harms are identified; can tend to inform the public about limitations inherent in the profession.	

Table 3.5.1. *Stakeholder Analysis: You discover you made an error and have to rerun everything.*

2. Identify decision-making frameworks

The ASA GLs reflect a virtue approach, which supports alerting stakeholders that the interpretations are incomplete and incorrect, because the ethical statistical practitioner "Avoids condoning or appearing to condone statistical, scientific, or professional misconduct. Encourages other practitioners to avoid misconduct or the appearance of misconduct." (H2), and also "Avoids compromising scientific validity for expediency" (E4). Although it would be more expedient to let the misleading interpretations stand, Principle B dictates that the ethical statistical practitioner: "Strives to promptly correct substantive errors discovered after publication or implementation. As appropriate, disseminates the correction publicly and/or to others relying on the results." (B5). Even though this erroneous report is only *based* on analyses and *correct* interpretations that you created, the correction is firmly your ethical and professional responsibility. The inconsistency of the final report with your original constitutes an error, as well as statistical, scientific, *and* professional misconduct *on the parts of those* who removed your caveats and present a misleading interpretation that can lead to erroneous decisions. *Their* misconduct is inconsistent with the ASA GLs; but your identification of the misconduct constitutes an ethical issue for *you* (A9).

The utilitarian perspective of the ACM CE means that the computing professional has a responsibility to assess the work they do/are asked to do so that decisions made on the basis of the system or its subparts are *justifiable*. If an error is identified in the use of correct evaluation (analysis) of a system, then results may be biased- as well as being incorrect, as was identified in this case - causing predictable but fixable harms. These harms are not *potential* but actual: once published or used in a decision, incorrect interpretations will be introduced into the literature (or into your organization's decision-making). One tragic example of this is the Space Shuttle Challenger explosion (https://en.wikipedia.org/wiki/Space_Shuttle_Challenger_disaster; see the Rogers Commission Report https://en.wikipedia.org/wiki/Rogers_Commission_Report where it is noted that the design flaw that caused the explosion was well known at least 9 years earlier). ACM CE 1.3 states that "A computing professional should be transparent and provide full disclosure of all pertinent system capabilities, limitations, and potential problems to the appropriate parties." It may be the case that, by reporting your correct interpretation as you did – to whomever you reported it to – was simply not sufficient: the people you thought were the "appropriate parties" actually did inappropriate things with your report. While your report itself was appropriate, the report*ing* was insufficient. In the vignette, you have identified a problem -not with *your own work*, but with the process by which your correct and transparent disclosure is communicated. This must be made clear – i.e., while you disclosed your caveats and concerns about the interpretability, it was not sufficiently "full" or transparent to ensure the propagation through to the final report. Note that, by following the organizational standard operating procedure and reporting to whomever you were directed to, you did not set out to create an error. Thus, as with the ASA GLs, the misconduct of those who misrepresent your report is inconsistent with the ACM CE; your identification of their misconduct constitutes an ethical issue for *you*.

3. Identify or recognize the ethical issue

Both the ASA and ACM standards support the identification and fixing of the errors of interpretation in the final report, with the ASA GLs stipulating that corrections should be disseminated (B5), and the ACM CE suggesting (implying) that the system of reporting that is in place may not be sufficient to prevent such misinterpretations, even when correct and full reports are actually made. While neither the GLs nor the CE specify that misconduct needs to be specifically identified, both are clear that the misconduct of others in misrepresenting your report creates the ethical issue for you, and so your responsibility is to ensure that the error of interpretation (rather than the misconduct itself) is identified and communicated to relevant stakeholders.

4. Identify and evaluate alternative actions (on the ethical issue)

The three decisions that can be made in any circumstance are: a) do nothing. b) consult or confer with a colleague (ASA) or a supervisor (ACM); and c) report violations of policy, procedure, ethical guidelines, or law (ASA) or refuse to implement the system (ACM).

As noted, there is only an ethical issue if you do not address the misinterpretation (e.g., publish or share/disseminate your complete report, with its correct interpretations) and notify all stakeholders that there was an error in the original report. Thus, option a) must be modified to "document your evidence that the prior interpretation is incomplete and misleading, notify all stakeholders, but do not confer"; and the only other option, b), becomes, "document your evidence that the prior interpretation is incomplete and misleading, and confer with peers or supervisors about how best to notify all stakeholders". If you identify and report the errors in the final report internally (following option b), and your supervisor(s) or organization then seek to suppress this information – particularly, by purposefully *not notifying clients or other stakeholders* who rely on the results, then option c) ("report the violation of CE, GLs, and relevant laws") is differentiated from option b) and the suppression of your correct results may actually be illegal as well as unethical.

5. Make and justify a decision

Decision: *options b and c* are the best supported decision in this case. The choice between b) and c) will depend on whether you encounter active suppression of your correct report. This case represents the heart of professional integrity and accountability (ASA Principle A), respect for the data and integrity of results (ASA B) as well as fulfilling responsibilities to stakeholders (ASA C), colleagues (ASA E), the profession (ASA F), and refusing to condone misconduct (ASA H2). Both b) and c) are also consistent with general ethical principles (ACM 1) and professional responsibilities (ACM 2).

6. Reflect on the decision

Reflecting on just how much of the ASA GLs are satisfied when you identify and correct your own errors can be reassuring: as noted in step 1, every one of the ASA GL Principles is informative about this case. This includes Principles H (the ethical statistical practitioner "Avoids condoning or appearing to condone statistical, scientific, or professional misconduct", H2). Because your report included all the correct caveats that contextualized your interpretation, and highlighted the potential for misinterpretation (by suggesting sensitivity analyses), but none of these are included in the final report, we may conclude that the misinterpretation errors were *not* honest ones. An ethical workplace (the focus of ACM 3 and ASA G, and Appendix) is supported when errors are identified and corrected because that is a recognizable characteristic of the ethical quantitative practitioner; purposeful misinterpretation is clearly statistical, scientific, and professional misconduct – and although this misconduct was exhibited by others and not you, you do bear responsibility to correct that or at least refuse to condone it. Whistle blowing may be in order (ACM 1).

The SHA shows that refusing to condone willful misconduct passes a *test of justice* -and doing nothing in this situation clearly fails that test. The SHA suggests that the harms that can accrue to the profession and the public trust – as well as to you, because the final report may be perceived as faulty because of your contributions (wholly or in part). The ACM caveat about consideration of whether whistle-blowing is warranted can benefit from consideration of the *test of publicity*: you would not want it publicized that you (incompetently) mis-interpreted your results -because this is clearly not the case here; however, if you failed to notify relevant stakeholders that your competent reporting and interpretations with caveats were ignored, then that *would* fail the test of publicity. By contrast, it *may* impact you negatively if it was publicized that your organization willfully (incompetently) *misrepresented* your results – and then failed to notify relevant stakeholders, and/or prevent you from notifying them, but in this case, the negative impact on you is much less than it is on your organization- or more specifically, whomever tried to suppress the correct interpretations of the results. There may be legal ramifications if it became known that your organization set out to purposefully mislead stakeholders by misrepresenting what was originally correct. The SHA shows the substantial harms (with minimal benefits) that accrue in this case if you did not take any of the options; doing nothing clearly fails the *test of universality*.

Questions for Discussion:

Do you agree with the decision in this case analysis? Why or why not? Which parts of the ER process are most and least acceptable?

Discuss the relevance to this case – and your analysis/decision – of policies regarding research misconduct, and policies specifically for *handling* misconduct. Are the Federal definitions (i.e., focusing on falsification, fraud, and plagiarism) helpful (or not)? How can you incorporate what you know about the differences between "misconduct" and "detrimental practices" into any part of the case analysis (prerequisite knowledge; identifying the ethical issue; determining alternative actions; making or justifying your decision; or reflection)? Note that the ASA mentions "misconduct" specifically, while the ACM does not. Discuss whether or not you feel that the ethical practice standards of these organizations actually have differing levels of relevance to discussions about misconduct and/or detrimental research practices?

Discuss the relevance to this case – and your analysis/decision – of policies in your organization regarding responsible authorship and publication. Are the ICMJE authorship criteria helpful (or not)? How can you incorporate ICMJE policies specifically into any part of the case analysis (prerequisite knowledge; identifying the ethical issue; determining alternative actions; making or justifying your decision; or reflection)? Do you feel that a policy featuring the ICMJE criteria for authorship could have helped in the reasoning or might be helpful in the future?

Keeping in mind that doing nothing/not responding are *not plausible responses*, are there other plausible alternatives (KSA 4) that you can think of? If not, discuss that; if so, list them and discuss their evaluation (are they equally consistent with GL/CE, do they lead to similar decisions, etc.).

Redo the analysis, but feature a different GL or CE principle in your reasoning process. Make sure the *justification* is different from what is given. Is the *decision* also different? Discuss your decision with its justification and the tests of universality, justice, and publicity.

Chapter 3.6
Documenting your work

Case: You are told <u>not</u> to document your work. When you do (because that's what the practice standards say the ethical practitioner does), your boss/supervisor returns it to you with the direction, "fix this".

1. Identify and 'quantify' prerequisite knowledge:

Which ASA/ACM Principles (and/or specific elements) seem most relevant to this vignette?

ASA:

Principle A: Professional integrity and accountability: A; A2, A9, A11, A12
Principle B. Integrity of data and methods: B1, B2, B3, B4, B6
Principle C. Responsibilities to stakeholders: C2, C4, C8
Principle E. Responsibilities to members of multidisciplinary teams: E2, E4
Principle F. Responsibilities to fellow statistical practitioners and the Profession: F4
Principle G. Responsibilities of leaders, supervisors, and mentors in statistical practice: G1, G5
Principle H. Responsibilities regarding potential misconduct: H2

ACM:

1. General Ethical Principles
2. Professional Responsibilities
3. Professional Leadership Principles
4. Compliance with the Code

Potential result: Stakeholder:	HARM	BENEFIT	UNKNOWN	UNKNOWABLE
YOU	Documentation takes time, adds effort (to fully document plans/evaluations/ systems); reporting this will result in inability to patent	Full documentation creates transparency and accountability, increasing reproducibility and documenting rigor.	Failure to document may lead to errors when others try to revise or reproduce your work.	
Your boss/client	Documentation takes extra time and may alert competitors to innovation or results before client is ready	Full documentation creates transparency and accountability	ACM: New harms/risks may be detectable more quickly (and addressed) with full documentation. ASA: replications are facilitated.	
Unknown individuals		Documentation creates transparency and accountability; open publishing of the method/ work increases access worldwide.		

Potential result:	HARM	BENEFIT	UNKNOWN	UNKNOWABLE
Stakeholder:				
Employer	Documenting takes time and may alert competitors to innovation/ results before client is ready; patents cost money even as they protect IP.	Demonstrates commitment to transparency & accountability; could add/add to a new patent/IP to improve organizational reputation	ACM: New harms/risks may be detectable more quickly (and addressed) with full documentation. ASA: replications are facilitated.	Harms may arise from loss of IP, if others implement this IP, it could limit business/profit
Colleagues	Documentation may add time (in reviewing) to colleagues' work. A patent would prevent them from using the IP themselves while documentation makes it fully accessible.	Full documentation creates transparency and accountability, and supports transparency, accountability, and replicability. Promotes peer evaluation.		

Potential result: Stakeholder:	HARM	BENEFIT	UNKNOWN	UNKNOWABLE
Profession	Documentation takes time, adds effort (to fully document plans/evaluations/ systems); publication of unprotected work may facilitate use of work that is unauthorized or does not acknowledge original creator; full documentation could possibly increase piracy	Documentation notifies the profession of improvements and innovations, and can move the field forward. Publication/ reporting increases visibility of the profession. Reports can simplify evaluations of the work (ASA) or system, and updates (ACM); add transparency and accountability. Promote and facilitates peer evaluation.		
Public/public trust	Without full documentation of work/thought processes, public trust in systems or their results, and the profession (ACM/ASA) are compromised. Perpetuating black box mentality does not help, and explicitly does not inform, society.	Transparency and accountability – and the potential to understand how decisions are made based on automation and systems - supports public trust in the profession and its work.		

Table 3.6.1. *Stakeholder Analysis: You are told not to document your work.*

2. Identify decision-making frameworks

Full and transparent documentation of your work is obviously an ethical obligation, but your *documentation is not the issue in this case*. It is the *prevention of your documentation by your boss* that creates the problem.

ASA GLs describe the ethical statistical practitioner as promoting reproducibility in all of their work; clearly you are trying to document fully and fulfill your ASA obligations from the virtue perspective, and behave as "the ethical statistical practitioner" does. The SHA supports transparent and complete documentation of your work, because harms may accrue when your work is not fully documented, and this also prevents others from comprehensively evaluating your work, creating additional harms. So, your documentation is clearly consistent with the utilitarian perspective of the ACM CE. While you are obliged to guard privileged information of your employer (ASA A11, B4), you also have responsibilities to assure your work is reproducible (ASA A2), and to guard against inappropriate methods (ASA B) and potential misleading of stakeholders (ASA C). Balancing these responsibilities can be difficult, so thinking through what the actual issue is (KSA 3) is valuable.

3. Identify or recognize the ethical issue

As this vignette is presented, you have applied your expertise to document your work appropriately, but someone else is trying to interfere in this ethical practice. *Your* actions – documenting your work – are clearly ethical, but your boss is telling you to violate the ethical practice standards, which is clearly unethical behavior *by that individual*.

4. Identify and evaluate alternative actions (on the ethical issue)

The same three decisions that can be made in any circumstance: a) do nothing b) consult or confer with a peer (ASA) or a supervisor (ACM); or c) report violations of policy, procedure, ethical guidelines, or law (ASA) or refuse to implement the system (ACM), are difficult to conceptualize when you have actually done the correct thing and someone is telling you to stop doing the ethical thing. Both practice standards require that you are honest and transparent in your documentation; both the ACM (Principle 4) and ASA (Purpose; A9, A12) outline specific responsibilities of the ethical practitioner to follow the practice standards *and encourage others to do so as well*. Also, ASA A11 outlines a responsibility to follow workplace policies *"unless there is a compelling ethical justification to do otherwise."* Similarly, ACM 1.2 specifies that "To minimize the possibility of indirectly or unintentionally harming others, computing professionals should follow generally accepted best practices *unless there is a compelling ethical reason to do otherwise."* (Emphasis in both added). There are virtue and utilitarian ethical reasons why not to follow this particular directive from your boss, if indeed this represents a workplace policy or somehow is a best practice. Thus, option a) changes from "do nothing" to "refuse to remove your documentation". Option b) becomes, "refuse to remove your documentation, and confer with peers or supervisors about how best to ensure that correct/full documentation becomes the standard operating procedure of your organization, *and* ensure that attempts to suppress transparent and complete documentation are recognized organization wide as unethical and not tolerated".

There may already be policies in place in your organization that should have prevented the attempted suppression of your documentation. If so, then option c) ("report the violation of CE, GLs, and relevant laws") *is* viable, and option c) effectively becomes, "refuse to remove your documentation, and confer with peers or supervisors about how best to ensure that correct/full documentation becomes the standard operating procedure of your organization, *and* ensure that attempts to suppress transparent and complete documentation are recognized organization wide as unethical and not tolerated, *and* report the cherry picking and suppression of its existence to the appropriate authorities (and to the ACM)." While the ACM does, but the ASA does not (as of May 2022), have a reporting mechanism, and a specifically articulated responsibility to report violations of the practice standards, there may be laws and other regulations that require full documentation. That means that the boss who tries to suppress your documentation is violating both ethical and legal practice standards.

5. Make and justify a decision

Decision: All options include "refuse to remove your documentation", because your original actions (including full documentation) are correct and ethically appropriate. Your boss may want you to act unethically, and if that is the case – particularly if that perspective is supported by your boss' bosses, then in addition to refusing to remove your documentation, you might also want to find other people to work with and for. In the case of the Cambridge Analytica scandal, if each person who was asked to behave unethically refused to do so, it is possible that at least one person could have been found at each step of the process, to comply – but all those who refused might have been noticed, or somehow able to notify the public earlier than March 2018, that the company was encouraging and maybe attempting to compel data scientists to behave unethically and illegally. The conferring (option b) and notification of authorities (option c) would have raised more awareness of the illegality and unethical activities, but for more junior practitioners, these can be very challenging steps to take.

6. Reflect on the decision

As with other cases, the practitioner has executed their job ethically but someone else's decisions to act unethically must be responded to. The ethical practitioner must be concerned with both their ethical practice and the context in which they practice, because if you continue without correcting others' misperceptions that you can be compelled to act unethically, you are actually violating multiple practice standards (and possibly, some laws). If others in your organization choose to ignore the ethical practice standards and relevant laws, you would need to decide if you wanted to work – and be associated with – that type of workplace and context. Consider how difficult it would be to get an interview for a job if you had been working at Cambridge Analytica throughout the period they were in the process of stealing data without consent. If you removed that from your resume, you would have a gap in employment; if you left it on, you might not be interviewed at all. If instead, you had it on the resume but also indicated that you left as soon as you noticed unethical/illegal behavior, and you reported that before leaving, it shows that you are both ethical and concerned about the wider community. The actions of the boss in this vignette to try and compel you to remove documentation – practice unethically – should be reported to their superiors, so that others will not be put into similar situations and also compelled to act unethically. This is what makes the notification features of options b) and c) so important. However, this can be a very uncomfortable thing to do. ASA GL Principle H offers guidance about how to conduct yourself when someone is accused of misconduct; ACM Principle 3 discusses how leaders should behave to promote ethical practice in the workplace. You might imagine that the more widely these two particular principles are known by statisticians and data scientists – and those who employ them – the better it could be when other practitioners feel the need to confer with colleagues or supervisors, or report it when they are coerced into violating ethical practice standards (or the law).

Sometimes, detrimental practices become part of the "norm", effectively training people to practice this way; –but sometimes people intend to purposefully mislead, which is the case for these vignettes, as they are based on steps in the Cambridge Analytica scandal. In business as well as research, it is unethical of you to withhold documentation; both the ASA and ACM standards also indicate that it is not ethical to engage in, condone, or promote such unethical behavior as "the norm".

The SHA suggests that all options pass a *test of justice*, because your refusal to behave unethically and withhold documentation strengthens the profession, and the public trust. All options pass the *test of publicity*: you *would* want it publicized that you are an ethical practitioner and fully document your work; however, your boss may not want that publicized – particularly if they act to repress your documentation. As noted in the previous case, an interest in keeping unethical behavior from being public is itself a *failure* of the test of publicity. All three options pass the *test of universality* as well, since you – and all stakeholders - definitely want all practitioners to be as transparent with documentation as you are in this case. While proprietary information of organizations must be guarded, and might not be reportable due to organizational policies on external communication, *internal* reports must be fully and transparently documented.

Questions for Discussion

Do you agree with the decision in this case analysis? Why or why not? Which parts of the ER process are most and least acceptable?

Discuss the relevance to this case – and your analysis/decision – of policies in your organization regarding mentor/mentee responsibilities and relationships. How can you incorporate what you know about these policies into any part of the case analysis (prerequisite knowledge; identifying the ethical issue; determining alternative actions; making or justifying your decision; or reflection)? If there is no policy in your organization relating to mentors, mentees, and their responsibilities in the mentor/mentee relationship, should there be one? Could such a policy have helped in the reasoning in this case, and/or do you think (in your reflection on the case and decision) such a policy is needed?

Discuss the relevance to this case – and your analysis/decision – of the role of peer review a) for your organization; and b) for your profession. Note that both GL and CE specifically mention the importance of peer review in strengthening your work (when you obtain it) and others' work, if you provide it competently. However, your organization may not have a policy relating to peer review (even though the practice standards mention that it is a key component of ethical practice). Do you feel that your organization should have some kind of policy about peer review? What do you think it should be/how should it be worded, and what would the justification for having one/not having one entail?

Keeping in mind that doing nothing/not responding are *not plausible responses*, are there other plausible alternatives (KSA 4) that you can think of? If not, discuss that; if so, list them and discuss their evaluation (are they equally consistent with GL/CE, do they lead to similar decisions, etc.).

Redo the analysis, but feature a different GL or CE principle in your reasoning process. Make sure the *justification* is different from what is given. Is the *decision* also different? Discuss your decision with its justification and the tests of universality, justice, and publicity.

Chapter 3.7
Reporting your results/communication

Case: You submit your complete and correct report of your scraping algorithm – including identification of the removal of your built in, opt-in consent to contribute data; the lack of consent accompanying data to be analyzed; and the lack of your recommendations in interpretations for limiting bias. You later discover that none of this documentation was included in the final report, but the final report is shared with stakeholders as if it is complete and correct.

1. **Identify and 'quantify' prerequisite knowledge:**

    Which ASA/ACM Principles (and/or specific elements) seem most relevant to this vignette?

ASA:

Principle A: Professional integrity and accountability: A6, A9, A11
Principle B. Integrity of data and methods: B5
Principle C. Responsibilities to stakeholders: C2, C4, C8
Principle D. Responsibilities to research subjects, data subjects, or those directly affected by statistical practices: D
Principle E. Responsibilities to members of multidisciplinary teams: E2, E3
Principle F. Responsibilities to fellow statistical practitioners and the Profession: F2
Principle G. Responsibilities of leaders, supervisors, and mentors in statistical practice: G1, G5
Principle H. Responsibilities regarding potential misconduct: H2

ACM:

1. General Ethical Principles
2. Professional Responsibilities
3. Professional Leadership Principles
4. Compliance with the Code

Potential result:	HARM	BENEFIT	UNKNOWN	UNKNOWABLE
Stakeholder:				
YOU	Incorrect and incomplete documentation means the work can no longer be used for valid or interpretable results. Your effort on the documentation has been wasted, *and* now neither the documentation nor what it describes can be used with confidence. Failing to correct the documentation may save time but *will* lead to unpredictable propagation of error throughout the workflow and into the community.	Correction of the misuse/editing means future work will be correct, reproducible, and interpretable.	Perpetuation of false reports and misinterpretation undermine ethical practice as well as any work based on the misleading documentation, while also perpetuating deceptive and detrimental research practices.	Deceptive and detrimental research practices like cherry-picking (what the editing in this case represents) undermine science, the public trust in the scientific enterprise, and the entire scientific community.

Potential result: Stakeholder:	HARM	BENEFIT	UNKNOWN	UNKNOWABLE
Your boss/client	Incorrect and incomplete documentation means the work can no longer be used for valid or interpretable results. All effort on the documentation has been wasted, *and* now neither the documentation nor what it describes can be used with confidence.	Correction of the misuse/editing means future work will be correct, reproducible, and interpretable.	Correction of this misuse of the documentation will require resources to fix/address. *Appropriate* documentation may not yield desired results.	
Unknown individuals	Incorrect interpretation undermines reproducibility and rigor. Misleading documentation negatively affects all stakeholders in the long run.	Correction of the error means future work will be correct, and valid.	Competent interpretation (ASA) supports reproducible results; effective contextualization of your work (ACM) promotes public good. Perpetuation of false reports and misinterpretation undermine both while also perpetuating detrimental research practices.	Deceptive and detrimental research practices like cherry-picking (what the editing in this case represents) undermine science, the public trust in the scientific enterprise, and the entire scientific community.

Potential result: Stakeholder:	HARM	BENEFIT	UNKNOWN	UNKNOWABLE
Employer	Incorrect and incomplete documentation means the work can no longer be used for valid or interpretable results. All effort on the documentation has been wasted, *and* now neither the documentation nor what it describes can be used with confidence. The incorrect documentation may be discovered to have been used by a client, or by other clients who paid for honest work, identifying past errors and undermining this and other projects	Correction of the documentation errors means future work with the product it describes (if not the employee who cherry-picked the documentation) may be facilitated –and correct, and valid.	Correction of the documentation errors may require resources to fix/address. *Appropriate* analysis –once the error is fixed - may not yield desired results.	Correction of the documentation errors may highlight past payments or purchases/sales of incorrect/error-prone work. Liability may be incurred.

Potential result:	HARM	BENEFIT	UNKNOWN	UNKNOWABLE
Stakeholder:				
Colleagues	Having to respond to this misuse of your documentation may complicate colleagues' work. Pointing out and correcting a team member's deceptive and detrimental research practices can be discomfiting – and may also highlight other detrimental practices that will then invite more scrutiny.	Identification and correction of the cherry picking allows mitigation of risks of harms (ACM); supports valid inferences (ASA), and can limit the unpredictable propagation of the error to other parts of the workflow. Other colleagues may be emboldened to call out – and stop – other deceptive and detrimental practices they observe.	Competent interpretation (ASA) supports reproducible results; effective contextualization of your work (ACM) promotes public good. Perpetuation of false reports and misinterpretation undermine both while also perpetuating detrimental research practices.	Deceptive and detrimental research practices like cherry picking (what the editing in this case represents) undermine science, the public trust in the scientific enterprise, and the entire scientific community.

Potential result: Stakeholder:	HARM	BENEFIT	UNKNOWN	UNKNOWABLE
Profession	Incorrect and incomplete documentation means the work can no longer be used for valid or interpretable results. If the documentation is shared, this will create invalidity in others' work. Failing to correct the documentation may save time but *will* lead to unpredictable propagation of error throughout the workflow and into the community.	Correction of the misuse/editing means future work will be correct, reproducible, and interpretable. Identification of the deceptive and detrimental research practice of cherry picking (what the editing in this case represents) may lead to more in the profession preventing it, and fewer people trying it.	Perpetuation of false reports and cherry picking undermine ethical practice as well as any work based on the misleading documentation, while also perpetuating deceptive and detrimental research practices.	Deceptive and detrimental research practices like cherry-picking (what the editing in this case represents) undermine science, the public trust in the scientific enterprise, and the entire scientific community.

Potential result:	HARM	BENEFIT	UNKNOWN	UNKNOWABLE
Stakeholder:				
Public/public trust	Deceptive and detrimental practices like cherry picking and misrepresentation undermine public trust in the plausibility of reproducibility and rigor – and, in the profession. An insistence (by the editor) on cherry-picked documentation being "correct" –resistance to its correction - undermines public trust as well as reproducibility and rigor generally. Without full documentation of work/thought processes, public trust in systems or their results, and the profession (ACM/ASA) are compromised. Perpetuating black box mentality does not help, and explicitly does not inform, society.	Identification and correction of the misleading documentation allows mitigation of risks of harms (ACM); supports valid inferences (ASA), and can limit the unpredictable propagation of the error to other parts of the workflow and community, to and decisions based on them. Transparency and accountability – and the potential to understand how decisions are made based on automation and systems - supports public trust in the profession and its work.	Competent interpretation (ASA) represents reproducible results; effective contextualization of your work (ACM) promotes public good. Perpetuation of false reports and misrepresentation in documentation undermine both while also perpetuating deceptive and detrimental research practices. Transparency supports trust, even if errors are discovered; their discoverability arises only and specifically from documentation, transparency, and accountability.	Deceptive and detrimental research practices like cherry-picking and willful misrepresentation undermine science, the public trust in the scientific enterprise, and the entire scientific community.

Table 3.7.1. *Stakeholder Analysis: Your report <u>was</u> complete and correct, but was edited to be incomplete and incorrect.*

2. Identify decision-making frameworks

This vignette describes a highly prevalent, deceptive, and extremely detrimental *research* practice, cherry picking[24]. Cherry picking is obviously not ethical, and in a research context, it occurs when results are selected because they fit a particular narrative or purpose. However, in this vignette, not only have results other aspects of your documentation been selected from the fuller report you provided without notifying the stakeholders, but the resulting report is also misleading – purposefully – to those stakeholders. This case is similar to the previous one about interpretation, but much worse: while you took the time and effort to fully and transparently report all of your work, someone in your organization, a team member, is ignoring – *suppressing* - the honest representation of your work and misleading the stakeholder readers. Because the editing that was done renders the documentation both incomplete and incorrect, its dissemination should be resisted, and corrected, by an ethical practitioner. Both the ASA and ACM practice standards state that honesty and trustworthiness are essential features of ethical practice; clearly the virtue perspective represents that the ethical practitioner is honest and trustworthy. The doctoring of the full report you provided is a violation of all stakeholder trust – including yours. This is detrimental not only to research and science, but to any enterprise where data, statistics, data science, and computing are involved.

"Ethical statistical practice supports valid and prudent decision making with appropriate methodology" (ASA Principle A), and the ethical statistical practitioner "avoids compromising scientific validity for expediency" (ASA E4). You obviously followed all the relevant ASA GLs, but at least one other person on the team seeks to subvert these ethical practices and misrepresent the work instead. That creates a new responsibility for you, A9, "Takes appropriate action when aware of deviations from these Guidelines by others." Additionally, Principle B dictates that the ethical statistical practitioner: "Strives to promptly correct substantive errors discovered after publication or implementation. As appropriate, disseminates the correction publicly and/or to others relying on the results." (B5) In addition to notifying stakeholders of the fact that the documentation is now incorrect – such that results arising from it will also be error-prone, possibly not reproducible, and probably invalid, the cherry-picking that the editing represents must be called out and recognized for the deceptive and detrimental practice that it is. The ethical practitioner must follow H2 and "Avoid condoning or appearing to condone statistical, scientific, or professional misconduct. Encourages other practitioners to avoid

---

[24] https://en.wikipedia.org/wiki/Cherry_picking

misconduct or the appearance of misconduct" – since the documentation that you originally produced was edited to be misleading, that clearly qualifies as all three types of misconduct.

The failure by the team member who did this editing to respect your original work represents statistical and scientific misconduct ("falsification"[25]), if not also professional misconduct. Problematically, the team member who did the editing is transmitting to others on the team, whether knowingly or not, some value for this misleading and detrimental work. Not only are they practicing unethically, they are, possibly subconsciously, encouraging others to do so as well (violating ASA G1, G5). The editing, and effort to mislead stakeholder readers, both perpetrates and condones statistical, scientific, and professional misconduct (ASA H1, H2). This vignette thus describes both a responsibility to correct the now-misleading report and also one to call attention to – and try to prevent – this transmission of deceptive and detrimental research practices by whomever did the revising of your documentation. This vignette describes a situation where the ASA GLs are, and should be, respected by someone who uses statistics and data science, even if they are not the ones who generated the results/report originally. The employer who supports this level of unethical practice – by encouraging or just failing to prevent or correct this behavior - violates all of ASA G (and the Appendix), and may actively be preventing ethical practitioners from following the other ASA GL principles. There may also be illegal behaviors at work, too, if the stakeholders who are misled are adversely affected by this falsification of the originally correct/complete report.

Although the ACM CE does not mention cherry picking or falsification, its utilitarian perspective states that "professionals should be cognizant of any serious negative consequences affecting any stakeholder that may result from poor quality work and should resist inducements to neglect this responsibility" (2.1). It also specifies that "a computing professional should be transparent and *provide full disclosure* of all pertinent system capabilities, limitations, and potential problems to the appropriate parties." (1.3, emphasis added). In this case, documentation that would support reproducible and interpretable, valid results, has been modified and the new (edited) documentation cannot be said to meet these standards - and that needs to be transparently reported. Deceptive and detrimental practice needs to be identified and discouraged at every possible opportunity. In the vignette, the documentation is incorrect and incomplete, so whatever results arise from the method or system the documentation describes will not function validly or reproducibly. This is a predictable limitation that will invariably cause harm – even if that harm is only

---

[25] https://ori.hhs.gov/definition-misconduct

to the scientific record. There is also harm in allowing the senior team member to influence the practice of others by demonstrating or suggesting that the sort of misrepresentation the editing creates in the final report is tolerable, violating ACM 3.1 and 3.4, as well as 4.1.

3. Identify or recognize the ethical issue

As this vignette is presented, the practitioner has applied their expertise and documented their work appropriately, providing a full and transparent report. But someone else has edited this work so that it is now incorrect and misleading. There is no way for the ethical practitioner to ignore this misuse (ASA A9), and allow this misrepresentation of their work to be shared with stakeholders (ASA B5). Both the ASA and ACM standards support reporting of the error, with both the ASA GLs and the ACM CE specifying that the errors (ASA) and limitations of the system (ACM, i.e., that it cannot yield valid results) should be disclosed. The ethical practitioner must respond, to correct the inappropriate editing and notify stakeholders that the final report is misleading and incomplete. The ethical issues arise: 1) if you *fail to notify* all stakeholders of the editing in the report, and also 2) if you do not make efforts *to identify and discourage the use of deceptive and detrimental practices*. The failure to notify is unethical because the misleading documentation will permit inappropriate use of the work (based on the now-inappropriate documentation), false and invalid reporting based on the incorrect use of the now-misleadingly documented system or method, and decisions that may be based on the incorrect documentation or the method/system it describes. All of these harms are predictable, and can be averted by notification. The ethical practitioner has a responsibility to call attention to cherry picking and other such misrepresentations of work as the unethical and deceptive, detrimental practice that it is, or else they violate ASA H2, failing in the virtue perspective.

In addition to notifying stakeholders of the misrepresentation of your work, there is a critical second ethical issue that the editor of the original report creates for you: because you recognize the deceptive and detrimental editing, you have an ethical obligation *to identify and discourage the use of deceptive and detrimental practices*. Both the ACM (Principle 4) and ASA (Preamble) outline specific responsibilities of the ethical practitioner to follow the practice standards *and encourage others to do so as well*. The knowledge that cherry picking is unethical does not always work to change behavior, but the ethical practitioner will *do what they can to discourage cherry picking and any other detrimental practices relating to statistics, computing, and data science*. To the extent possible given your role (and job security), an ethical issue for the practitioner arises when they do not seek to discourage this deceptive and detrimental behavior. The harms that accrue when incorrect and incomplete documentation are permitted – especially when this is the result of misuse of appropriate documentation – violate elements of *every* ASA GL and ACM CE principle.

4. Identify and evaluate alternative actions (on the ethical issue)

As usual, the three decisions that we discuss for any circumstance are: a) do nothing. b) consult or confer with a peer (ASA) or a supervisor (ACM); and c) report violations of policy, procedure, ethical guidelines, or law (ASA) or refuse to implement the system (ACM). The practitioner was ethical in their original documentation, but to ignore the editing, and "do nothing", is clearly in violation of ethical practice standards in terms of both of the ethical issues that this vignette identifies. Option a) should change to "correct the erroneous documentation, notify all stakeholders that the original documentation was edited to be incorrect and incomplete, and act to prevent the use or further publication of the false documentation." Option b) becomes, "confer with peers or supervisors about how best to correct the erroneous documentation, notify all stakeholders that the original documentation was edited to be incorrect and incomplete, and act to prevent the use or further publication of the false documentation, and ensure that cherry picking and other detrimental practices (including suppression of evidence of cherry picking or spurious results) are recognized as unethical and not tolerated".

The 2017 National Academies of Sciences and Engineering and the Institute of Medicine (NASEM) report, "Fostering Integrity", called on the scientific community in the US to identify and act to prevent deceptive and detrimental practices such as cherry picking. In this vignette, you fully and transparently documented your work. There may be a clear policy in your workplace/organization that should have prevented the falsification of your documentation. So, option c) ("report the violation of CE, GLs, and relevant laws") is a viable option. While cherry picking is unethical but not illegal, deception and withholding evidence is illegal – particularly when there are financial conflicts of interest[26] in play. If the senior team member seeks to suppress the correction of misleading and incorrect documentation, they may be perpetrating fraud[27] (while also engaging in, and implicitly or explicitly encouraging detrimental and deceptive practices). Both your commitment to ethical practice and your civic duty will require that you report this. Then option c) effectively becomes, "correct the erroneous documentation, and notify all stakeholders that the original documentation was edited to be

---

[26] https://ori.hhs.gov/education/products/ucla/chapter4/default.htm
[27] https://en.wikipedia.org/wiki/Fraud

incorrect and incomplete, and act to prevent the use or further publication of the false documentation, and ensure that cherry picking and other detrimental practices (including suppression of evidence of cherry picking or spurious results) are recognized as unethical and not tolerated, and report the cherry picking and suppression of its existence to the appropriate authorities (and to the ACM)." Note that the ACM does, but the ASA does not (as of May 2022), have a reporting mechanism as well as a specifically articulated responsibility to report violations of the practice standards. Note also that, while the NASEM report and a lot of discussion are around research misconduct, these violations are also important outside of scientific applications of statistics, data science, and computing.

5. Make and justify a decision

Decision: All options include "correct the erroneous documentation, notify all stakeholders that the original documentation was edited to be incorrect and incomplete, and act to prevent the use or further publication of the false documentation". These are the only ethical responses – and there are clearly two parts to the response - to the unethical, deceptive, and detrimental behavior in this vignette. The differences across the options are in how the response is made, which may depend on features of your specific context. Options b and c include sharing (i.e., reporting, option c) or conferring (option b) with others. Option c) "reporting" should be considered, particularly if this is not the first experience you are having (or that you know of) where team members have misrepresented and/or falsified documentation. Consulting a peer or supervisor as to how best to proceed (option b) may be more relevant for a more junior practitioner. As you grow in experience and seniority, you may be able to use the conferring (option b) to give some careful thought given to why no one else on the team objected to the reporting of cherry-picking/editing of your documentation, and to how others are trained and mentored in their reporting at your organization. As you assume leadership roles through your career, evaluating the origins of unethical or detrimental practices will help you to follow G1 and G5, and to avoid both E4 and H2.

6. Reflect on the decision

The SHA shows that in this case, the main harms ("major", highlighted in dark grey) that failures in this case can lead to accrue to you and to the profession: specifically, these harms are in the form of permitting misuse, abuse, and potentially illegal misrepresentation of your work. The ASA and ACM ethical practice standards both clearly charge the ethical practitioner with the identification of errors and notification of stakeholders, but the potential misuse of your documentation is not just "unethical", it might also be illegal –

and of course, people who are willing to break the law will not care that they are also violating ethical practice standards. However, both practice standards also clearly charge the ethical practitioner with doing what they can to ensure that others follow the standards; they are also explicit about the fact that even those who are not "statisticians", "data scientists", or "computing professionals" full time but who do use the methods and technology are obliged to follow the ethical guidelines.

There may be a culture in your team or organization that either tolerates or encourages detrimental practices, and the ethical practitioner is obliged to do what they can change this culture; "what you can do to change the culture from one that supports or even encourages detrimental practices to an ethical one" will naturally change with seniority and experience – but it will only become more important over time, and will never be less important! When an otherwise ethical practitioner fails to stop others' unethical practice, they would then violate ASA H2 and "condone, or appear to condone", these unethical practices – meaning that a team or organizational culture or context where you are encouraged or directed to violate your practice guidelines is an unethical culture. Options b) and c) are preferred over option a) because the notification aspect of these options may prompt organization-level review of policies or incentives that exist for reporting unethically. Option b) may be more appropriate to initiate culture change, and to lead those who are ignorant that their behavior is unethical (e.g., if their practice guidelines do not decry deceptive and detrimental practices) to the understanding that cherry picking is unethical. However, option c) may be the decision that has the greatest impact, preventing frank deception from propagating from your work, and also promoting ethical practice by alerting the organization to the senior member's pattern of deceptive and detrimental practice.

Note that, once you have identified an error in the final report, if you continue without correcting it (even though this is the easiest thing to do), you are violating multiple practice standards. If others choose to continue using – and making decisions based on - a report in which you have identified errors, and they purposely ignore the errors, they are also violating multiple practice standards. You would have to decide if you wanted to work – and be associated with – that type of workplace and context.

Sometimes, detrimental research practices are engaged in by people who were trained to practice this way –but sometimes they are engaged in to purposefully mislead, which is the case for these vignettes, as they are based on steps in the Cambridge Analytica scandal. In business as well as research, it is unethical of you to mislead anyone into thinking that results are correct and valid; it is unethical to knowingly contribute false, irreproducible, and/or invalid results and inappropriate methods to the literature. A failure to correct this error and remain an author on the publication means that you endorse making this irreproducible and invalid "contribution" to the literature.

The SHA shows plainly that all options pass a *test of justice*, because your notification of others strengthens the profession and the public trust while correcting an error that was inadvertently contributed to the literature. All options pass the *test of publicity*: you *would* want it publicized that you are an ethical practitioner, and that is what the correction and stakeholder notification will achieve; however, your supervisor or organization may not want that publicized. Keep in mind that an interest in keeping unethical behavior from being public is itself a *failure* of the test of publicity! The three options pass the *test of universality* as well, since you – and all stakeholders - definitely want all practitioners to be as vigilant as you are in this case.

Questions for Discussion

Do you agree with the decision in this case analysis? Why or why not? Which parts of the ER process are most and least acceptable?

Discuss the relevance to this case – and your analysis/decision – of policies regarding conflicts of interest in your organization. How can you incorporate what you know about these policies into any part of the case analysis (prerequisite knowledge; identifying the ethical issue; determining alternative actions; making or justifying your decision; or reflection)? Note that, if there is no alignment between such policies and this case, you can comment on that (e.g., do policies in your organization focus on *avoiding* COI or *managing* them?). Do you feel that these policies could have helped in the reasoning?

Discuss the relevance to this case – and your analysis/decision – of policies in your organization regarding responsible authorship and publication. Are the ICMJE authorship criteria helpful (or not)? How can you incorporate ICMJE policies specifically into any part of the case analysis (prerequisite knowledge; identifying the ethical issue; determining alternative actions; making or justifying your decision; or reflection)? Do you feel that a policy featuring the ICMJE criteria for authorship could have helped in the reasoning or might be helpful in the future?

Keeping in mind that doing nothing/not responding are *not plausible responses*, are there other plausible alternatives (KSA 4) that you can think of? If not, discuss that; if so, list them and discuss their evaluation (are they equally consistent with GL/CE, do they lead to similar decisions, etc.).

Redo the analysis, but feature a different GL or CE principle in your reasoning process. Make sure the *justification* is different from what is given. Is the *decision* also different? Discuss your decision with its justification and the tests of universality, justice, and publicity.

Chapter 3.8
Engaging in team science/work

Case: Leadership informs your team that they bought an algorithm that you will be using to scrape data. But first, they want you to take off all the consent pop-ups, because "that ruins the user experience" and "adds personal data we will only need to strip off to preserve confidentiality".

1. Identify and 'quantify' prerequisite knowledge:

Which ASA/ACM Principles (and/or specific elements) seem most relevant to this vignette?

ASA:

Principle A: Professional integrity and accountability: A3, A9, A11
Principle B. Integrity of data and methods: B1
Principle C. Responsibilities to stakeholders: C7
Principle D. Responsibilities to research subjects, data subjects, or those directly affected by statistical practices: D; D1, D4, D5, D10, D11
Principle E. Responsibilities to members of multidisciplinary teams: E2, E4
Principle G. Responsibilities of leaders, supervisors, and mentors in statistical practice: G1, G5
Principle H. Responsibilities regarding potential misconduct: H1, H2
Appendix: Responsibilities of organizations/institutions

ACM:

1. General Ethical Principles
2. Professional Responsibilities
3. Professional Leadership Principles
4. Compliance with the Code

Potential result: Stakeholder:	HARM	BENEFIT	UNKNOWN	UNKNOWABLE
YOU	Using data with unknown provenance, particularly if some of it is known not to have been contributed with informed consent, violates the data contributor rights as well as ethical practice standards.	While a violation of multiple specific elements of the professional practice standards and some laws, the sampling might be less biased/more representative if everyone is 'forced' to contribute data.	Failing to obtain consent to use data and using stolen data could lead to unpredicted harms, bias, or unfair results, as well as risks to data contributors – with the statistician or data scientist bearing responsibility for misuse, unauthorized access, or losses of that data.	Using stolen data and data that was not contributed with informed consent may suggest to others/other system developers that collecting stolen/breached data is OK, even though this directly violates practice standards.
Your boss/client	Data with unknown provenance can lead to unpredicted harms, bias, or unfair results, and can create risks for the data contributors that were foreseeable.	Ignoring the provenance of data, while a violation of multiple specific elements of the professional practice standards, is simpler and cheaper than ensuring data are obtained with proper consent.		

Potential result: Stakeholder:	HARM	BENEFIT	UNKNOWN	UNKNOWABLE
Unknown individuals	Data that is stolen or accessed without authorization and consent (i.e., from breaches) may expose data contributors to risks, as well as to (further) misuse by others.	There are no benefits that accrue when stolen data are utilized, and no one stops the use of that kind of data.		Using stolen data and data that was not contributed with informed consent may suggest to others/other system developers that collecting stolen/breached data is OK, even though this directly violates practice standards.
Employer	Using data with unknown provenance could lead to unpredicted harms, bias, or unfair results, and can create risks for the data contributors that were foreseeable, thus incurring liability to the employer.	Ignoring whether individuals do or do not consent to contribute their data, while a violation of multiple specific elements of the professional practice standards, may lead to a more representative sample than if only data obtained with proper consent are used.		

Potential result: Stakeholder:	HARM	BENEFIT	UNKNOWN	UNKNOWABLE
Colleagues	If others on the team are not aware that the data provenance is mixed (and some is obtained even though the contributor did not consent to give it), colleagues may mistakenly share –i.e., further the misuse of- the data.	While a violation of multiple specific elements of the professional practice standards and some laws, the sampling might be less biased/more representative if everyone is 'forced' to contribute data.		
Profession	The profession may appear untrustworthy when it is discovered that practitioners created and then used a method to ensure they could take your data even if no consent was given. Using data with unknown provenance is a violation of multiple specific elements of the professional practice standards.	There are no benefits that accrue when methods of obtaining informed consent are circumvented, and no one stops the use of that kind of data.	Failing to obtain consent to use data and using data obtained without consent could lead to unpredicted harms, bias, or unfair results, as well as risks to data contributors - with the statistician or data scientist bearing responsibility for misuse, unauthorized access, or losses of that data.	Using data that was not contributed with informed consent may suggest to others/other system developers that collecting stolen/breached data is OK, even though this directly violates practice standards. This decrements professional integrity in a concrete way.

Potential result:	HARM	BENEFIT	UNKNOWN	UNKNOWABLE
Stakeholder:				
Public/public trust	Public sentiment about the security of their data will continue to worsen, and people will become less inclined to contribute or sharing data if public concerns about data security, or the lack of honesty in data collection systems, continue.	There are no benefits that accrue when stolen data are utilized, and no one stops the use of that kind of data.		

Table 3.8.1. *Stakeholder Analysis template: Leadership buys an algorithm and tells you to remove any consent feature when it is running.*

2. Identify decision-making frameworks

The ethical statistical practitioner has responsibilities to "Understands and conforms to confidentiality requirements for data collection, release, and dissemination and any restrictions on its use established by the data provider (to the extent legally required)..." (C7). Moreover, the ethical statistical practitioner "Uses data only as permitted by data subjects' consent when applicable or considering their interests and welfare when consent is not required. This includes primary and secondary uses, use of repurposed data, sharing data, and linking data with additional data sets." (D5). Thus, utilizing data with unknown provenance violates these – and other -ASA GLs. Moreover, circumventing a mechanism whereby humans are given a choice to contribute and say no is frankly unethical and may be illegal. Obviously, failing to protect basic human rights by using data you do not have consent to use violates both C7 and D5, but also violates ASA GL Principle H2 ("Avoids condoning or appearing to condone statistical, scientific, or professional misconduct. Encourages other practitioners to avoid misconduct or the appearance of misconduct") as well as federal and international laws in some cases. In all of these violations, the virtue perspective is clear: this is unethical for a diverse range of reasons.

Not all statistical, computing, and data science applications will end up in the research record – in those cases, the definitions of *scientific misconduct* are less relevant and taking people's data even though they choose not to give it is clearly *statistical* and *professional* misconduct. The specific directions to remove the consent features is a clear signal, by leadership, that consent might actually be obtainable, but the choice of data subjects to opt out is not wanted by the leaders of your organization. This is completely inconsistent with the virtue perspective of the ASA GLs, and constitutes a "compelling ethical justification" to do otherwise that follow these directions (A11).

The ACM CE, with its utilitarian perspective, focuses on limiting harms and in this case, there are many harms, and no benefits, to using data that was contributed to your analysis without consent; the harms are all compounded – and with no benefits –when the choices and exercise of basic human rights not to contribute data are circumvented. Moreover, a system that collects data without obtaining consent may include methodologies or other features that are insecure or otherwise create risks (in addition to risks of confidentiality and privacy breaches) that are not addressed – because the manner in which data are collected may not be legitimate or sufficiently specified. In this case, you simply do not know what the provenance will be of the data that this purchased algorithm collects; some contributors may have consented, and others would not have. The method for circumventing non-consenters to take their data anyway might be characterized by leaders as a "strength" of the algorithm they purchased, but it is clearly a violation of basic human rights and principles of autonomy, offsetting the technical strength with profound cultural, social, legal, and ethical *limitations* (i.e., it fails to follow ACM CE Principles and to limit harms/prioritize the public good). Thus, the utilitarian perspective also permits a "compelling ethical justification" to do otherwise that follow these directions.

3. Identify or recognize the ethical issue

Both the ASA and ACM practice standards state explicitly that professionals must respect laws and practice standards to ensure data are collected with the knowledge and consent of contributors. The primary ethical issue in this case is that both GLs and CE require that data be obtained with consent and the contributors' knowledge of what the data will be used for, but the organization leadership is instructing you not to do this. Since it is patently unethical to use data that was not obtained with informed consent, it is clearly unethical to pretend that you or your system respects their choice not to consent, and also unethical to modify an algorithm to remove a consent feature that exists. Thus, there is one ethical issue: you are being instructed by leadership to do something that violates ethical practice standards. However, your response(s) comprise multiple aspects of ethical practice of statistics and data science.

4. Identify and evaluate alternative actions (on the ethical issue)

As usual, the three decisions that can be made in any circumstance are: a) do nothing; b) consult or confer with a peer (ASA) or a supervisor (ACM); and c) report violations of policy, procedure, ethical guidelines, or law (ASA) or refuse to implement the system (ACM). Clearly, "do nothing" is both totally inappropriate in this case, and also a violation of the practice standards (and laws) relating to respecting data contributors. As alluded to in step 3 above, the response to the ethical issue actually features multiple dimensions of ethical practice. Firstly, the instructions cannot be followed: this is possibly the only case where "do nothing" is actually viable! But of course, instead of "do nothing" (which is technically not an ethical response), the first dimension is to refuse to follow these instructions. In addition, you have an ethical obligation to stop the use of any algorithm that circumvents non-consent, because ASA H2 ("Avoids condoning or appearing to condone statistical, scientific, or professional misconduct. Encourages other practitioners to avoid misconduct or the appearance of misconduct.") and ACM 2.1 ("Computing professionals should insist on and support high quality work from themselves and from colleagues.") and 4.1 ("Computing professionals should adhere to the principles of the Code and contribute to improving them. Computing professionals who recognize breaches of the Code should take actions to resolve the ethical issues they recognize, including, when reasonable, expressing their concern to the person or persons thought to be violating the Code.") require that you not condone unethical practice by others, and anyone who does follow these instructions will clearly be acting unethically. Finally, in the event that someone does follow these unethical instructions, there may be a third dimension to consider: do not proceed with any use of the data except what was actually contributed with informed consent.

Thus, we can change option a) from "do nothing" to "refuse to implement the consent feature removal, *and* stop the use of any algorithm that circumvents non-consent, *and* do not proceed with any use of the data except what was actually contributed with informed consent". Then option b) would become, "refuse to implement the consent feature removal, *and* stop any use of the data except what was actually contributed with informed consent, and consult or confer with a peer (ASA) or a supervisor (ACM) as to how best to stop the use of the method that circumvents non-consent". Option c) must also be modified, because simply reporting the fact that the proposed modifications to the data collection algorithm violates the GLs (ASA) and basic human rights, or refusing to implement the modified system (that collects data when consent is not given) (ACM) might not ensure that the data with unknown provenance is both not used and also, that its collection ceases. So, option c) needs to change to "report violations of policy, procedure, ethical guidelines, or law to a peer (ASA) or a supervisor (ACM), *and* stop the collection of non-consented data, *and* stop any use of the data except what was actually contributed with informed consent."

5. Make and justify a decision

Decision: *Note that all options include* "refuse to implement the consent feature removal, *and* stop the use of the method that circumvents non-consent, *and* do not proceed with any use of the data except what was actually contributed with informed consent" – which is consistent with what we discussed for KSA 4. What else is done (nothing else in option a); *conferring* on how best to stop the method in option b); and not conferring but reporting to relevant authorities the instructions to modify the algorithm, and potential existence of a data scraping algorithm that is lacking any consenting features (particularly when it had such a feature to begin with) in option c) may depend on your level of seniority.

Clearly the non-consented data collection – as well as its use – must stop, because both are unethical. If data with unknown provenance is commonly collected in your work context, it might be impossible for you choose even option a) and stop the mechanism as well as all use of data with unknown provenance. In that case, it is impossible for you to do *anything* ethical except notify relevant stakeholders and authorities of the situation, and then seek another team, other colleagues, or a different job.

Option b) would be the *most desirable option* in that consulting with a colleague/supervisor will ensure at least some kind of notification that, or publicity for the fact that, data collection without consent is frankly unethical, as is the use of such inappropriately obtained data. There may not be policies in your workplace against such interference/violation of the ASA GLs or ACM CE, so reporting (option c) may not be a viable option, making it a slightly less desirable option than b. However, in contexts where such policies do exist, it may still be difficult to report such behaviors; but you (the practitioner) can – and should - still demonstrate your professional competence and follow the relevant practice standards. Option c may not be as much about "reporting" a person or practice as it is about "educating" people or your company/ organization about the practice standards. It is important *to the profession* to support the next practitioner who ends up in a similar situation (or better, to prevent such unethical data collection in the future), so it is definitely worth considering politely notifying relevant individuals that the data collection system violates the ACM CE, and you cannot be directed to violate the CE and GLs. When you present the system and it features modifications that may limit the amount of data but also limit the likelihood of including stolen/breached data, it would be an opportunity to point out to everyone involved –and everyone in attendance - that this is the only way to meet ASA GLs/ACM CE and limit harms/risks of harms.

6. Reflect on the decision

Seniority of staff can complicate your efforts to follow ethical practice standards; if they are your boss or supervisor, or friends with your boss/supervisor, it can seem daunting to report to them that their friend or close colleague is behaving in an unprofessional (and unethical) way. While not specified, the ACM preamble does state that "(t)he entire computing profession benefits when the ethical decision-making process is accountable to and transparent to all stakeholders. Open discussions about ethical issues promote this accountability and transparency." It will be supremely uncomfortable to initiate a conversation about unacceptable behavior in the workplace – especially when it is exhibited by leaders; however, it is essential to keep in mind that doing nothing is also unacceptable. In Chapter 3.9 we will discuss how important it is to consider exactly how you will go about initiating, and then responding within, a conversation around an ethical challenge and/or the response you propose that will address that ethical issue. Like all the CE and GLs, there is no algorithm or rule about how to go about initiating, and then completing, conversations that may be uncomfortable.

In this vignette, however, the violation of the practice standards and applicable laws, and possibly also policies in your workplace, are clear. It should also be clear that *to do nothing is unethical* and *unacceptable*. The leaders clearly purchased an algorithm that did include a mechanism that seeks consent for data contribution. The instruction to remove that feature prioritizes the implementation of the algorithm over human rights to autonomy and the public good. The rationales accompanying the instruction are obviously false, since if giving consent truly "ruins the user experience", then it means those users would likely *not* be consenting – which should be their right, and that opportunity to decide should be given, and the decision should be respected. The suggestion that consent (or not consent) "adds personal data we will only need to strip off to preserve confidentiality" is more reflective of an inadequate understanding of how to implement the algorithm – i.e., whomever suggests that consent represents "personal data" – and especially that indication of a person's choice about their data contribution (i.e., consent) should be stripped off, and not used, is totally incorrect. They advocate the varieties of ethical violations outlined above, but they themselves are violating the ethical practice standards articulated by both the ACM (2.6) and ASA (A) that the ethical practitioner only accepts jobs that they are competent to execute. A person who argues consent will ruin a user's experience and generate data that is personal and in need of removal is clearly not a competent practitioner.

Complying with the leadership request creates harms to all stakeholders, including tricking users of the data into believing that the data were *collected with consent*. An individual who removes, or directs employees to remove, a consent feature is seeking to circumvent refusals to give consent. Such an individual might also be unlikely to want this particular feature fully documented, since it reflects frankly unethical data collection. Thus, in addition to instructing employees to violate both ethical practice standards and applicable laws, the leadership in this vignette seeks to violate most ethical principles in the ACM and ASA standards, and to encourage their employees to do so as well. No legitimate benefits –certainly none that can offset the harms to all stakeholders – can accrue, so the utilitarian perspective as well as the ACM CE point out that this request is clearly not ethical. Complying with these instructions will create additional violations of human rights, as well as of ACM Principles – causing all those who use the data from this dishonest system to violate human rights and other CE Principles as well. Finally, complying with this directive could easily create or perpetuate unfair outcomes, violating ASA A3.

Each of the alternative responses to the vignette passes the tests of publicity, universality, and justice. Although it might seem counter to your career trajectory to let it be known that you blew the whistle on frankly unethical leadership in the workplace, the test of publicity would instead let it be known that you do not condone or tolerate unethical behavior and professional misconduct (and that is an excellent thing to publicize!). If everyone in the workplace either supported or helped in the documentation of misconduct like this (i.e., option a), it would eventually become impossible for leaders to disrespect both data contributors and the potential users of this type of data in this way; and reporting such misconduct might result more directly in removing such unethical individuals from the workplace entirely. Therefore, all options do pass the test of universality. In addition to taking direct action to stop unethical and potentially illegal behavior like that of the leaders in this vignette in the workplace, all of the options tend to create a more uniformly respectful workplace for all – so all options also pass the test of justice.

Questions for Discussion

Do you agree with the decision in this case analysis? Why or why not? Which parts of the ER process are most and least acceptable?

Discuss the relevance to this case – and your analysis/decision – of policies regarding research misconduct, and policies specifically for *handling* misconduct. Are the Federal definitions (i.e., focusing on falsification, fraud, and plagiarism) helpful (or not)? How can you incorporate what you know about the differences between "misconduct" and "detrimental practices" into any part of the case analysis (prerequisite knowledge; identifying the ethical issue; determining alternative actions; making or justifying your decision; or reflection)? Note that the ASA mentions "misconduct" specifically, while the ACM does not. Discuss whether or not you feel that the ethical practice standards of these organizations actually have differing levels of relevance to discussions about bullying, and whether or not this is professional misconduct, a detrimental practice, or something else?

Discuss whether this case represents an opportunity for those involved to reflect on "the scientist or practitioner as a responsible member of society". If so, is it limited to "society" defined as the community of practitioners in the same field (i.e., to "the profession"), or is the case also (or only) relevant to the scientist or practitioner as a responsible member of the wider society?

Keeping in mind that doing nothing/not responding are *not plausible responses,* are there other plausible alternatives (KSA 4) that you can think of? If not, discuss that; if so, list them and discuss their evaluation (are they equally consistent with GL/CE, do they lead to similar decisions, etc.).

Redo the analysis, but feature a different GL or CE principle in your reasoning process. Make sure the *justification* is different from what is given. Is the *decision* also different? Discuss your decision with its justification and the tests of universality, justice, and publicity.

Chapter 3.9
Embracing your inner ethical practitioner: engaging in open conversations

The reader is invited to return to each of the chapters and case vignettes in Section 3. Each case analysis that is provided (as well as those analyses that readers conduct themselves, according to the discussion questions in each chapter), should be revisited through the lens of *role playing*. Specifically, readers are invited to practice – in actual role-playing dyads, or in writing, or both – the delivery of their ethically-reasoned decision in the workplace.

Attention should focus on not just the structure of the case analysis, but how to broach the subject with the peers. If the reader's workplace has anyone in leadership roles that have an "open door policy", then readers should imagine role playing (or just contemplating) how to engage that leader in any of the discussions the Section 3 case analyses outline. Part of the alternative decision of conferring with a peer could be determining how best to approach leadership on what the ethical challenge is, and how to address it. Some of the cases, as discussed in each chapter, are more complicated than others. Role playing, whether you actually do it or just imagine how it would go, could facilitate a conversation with someone in a leadership position might positively impact the ethical work climate – e.g., enabling those in leadership, supervisory, or mentorship roles to fulfill obligations under ASA Principle G and/or ACM Principle 3.

Readers may feel more confident in engaging in conversations with others if they are prepared with a full, written analysis of what the issue is/issues are, what alternatives were considered, what decision is recommended/was taken and why, and some reflection on the relevance of the decision. The formal analysis is a process that may require some practice, and will almost certainly require familiarity with both the appropriate ethical practice standard(s) – i.e., from the ASA or ACM – and also contextually-appropriate policy or other guidance. For example, a practitioner in the US Federal context, like the Bureau of Labor Statistics or US Census Department, would utilize the current version of *Principles and Practices for a Federal Statistical Agency* (National Academies of Sciences, Engineering, and Medicine, 2021; in its 7th edition) as well as the ASA or ACM practice standards. These guidelines all agree on what is ethical behavior/how to practice ethically. But the individual with whom you are discussing a case, a case analysis, or how to identify, and make a decision about, an ethical challenge will want or need to see the guidance in order to better understand the situation, potential solutions, and/or how to share these. For these reasons and more, following the ethical reasoning KSAs will not only help you to write up what the issues are and why they are issues, but it will also help you figure out at least some options. If you're able to take advantage of a leader's open-door policy, maybe that leader will have additional options to discuss or consider.

The reason why a whole chapter is dedicated to role playing is that engaging in these kinds of conversation requires some practice! The case analysis, following the ethical reasoning KSAs, and utilizing the ASA or ACM practice standards – and/or any other workplace policy guidance – is what Section 3 was designed to help you practice. Utilizing those case analyses in conversations that can lead to a more ethical workplace is the natural next step – but also requires attention and practice. Beginning with the case analyses that have already been laid out in Section 3, but then moving on to the readers' own analyses and critiques, role playing will help readers to understand the dynamics that might not have been as clear when working through ASA Principles E, F, and G or ACM Principle 3.

Whatever the source of the cases, readers should role play different sides of the conversation:

-   The case analyzer role: Describe your analysis of the case and your decision. Are you asking for confirmation of your conclusions/decision, or are you asking for assistance in reaching that decision?
-   The case co-analyzer role: You "receive" a case analysis, and a peer or co-working asks you to confirm the analysis, and collaboratively determine the best way to ensure such situations do not recur.
-   The preventer of ethical case analysis role: You receive a case analysis, but you don't want it to be formalized or shared. You try to dissuade the analyzer from making – or publicizing – that decision. You don't know whether they've done the analysis right – you just don't want anyone to rock the boat.
-   The responder-to-preventer role: You or a peer have completed a case analysis, and you shared it with a supervisor or team leader. This person does not support your analysis and your decision. They do not want you to rock the boat.

With respect to each of these responses to your case analysis, readers should discuss or consider how the ASA Guidelines and ACM Code of Ethics promote professional identity, and/or, professionalism (in you). The rationale for adding professional identity and professionalism to the readers' engagement is to reinforce the importance of these topics; when taking action and having or starting a conversation might be uncomfortable, recognizing that the uncomfortable conversation could have a beneficial result or outcome can be a great motivator.

Begin with choosing the case from Section 3 that you feel would be the most uncomfortable, and/or the most difficult conversation to initiate. Consider why it would be uncomfortable and difficult. This might be due to a desire to avoid confrontation; a sense of "I'm too junior to fully understand or complain about such activities"; or an unacknowledged recognition that, if this kind of behavior is going on at your workplace, then you'd much rather just get a different job than go through the difficult and uncomfortable experience of addressing the problem. Changing jobs because the workplace setting is too unethical is an excellent strategy – one day, such workplaces will simply have to change their culture because no competent practitioners will want to work there! However, just leaving without trying to strengthen the culture of that workplace against unethical, detrimental, or illegal practice is akin to "doing nothing". This is likely to be the easiest approach, but it is not consistent with the ASA or ACM practice standards, as you have seen. Hopefully, considering why the conversation would be hard/uncomfortable will help it be less problematic; at the very least, with a formal write up of an ethical violation (like those in the preceding chapters of Section 3), you would be able to deliver a detailed report to the Human Resources department after securing your next job.

Chapter 3.10

Summary of Section 3 and the book: career spanning engagement in professional and ethical practice

In each chapter of Section 3, we analyzed one vignette per task and readers were invited to analyze that case analysis. A few aspects of the importance of ethical reasoning, and the ethical practice standards of the ASA and ACM, should be noted. Firstly, as the vignettes in Section 3 moved along the statistics and data science pipeline, it might seem like more and more of the ethical practice standards became relevant to devising a solution. You may recall from Section 2, where there were no ethical challenges (so, no solutions), much of the ethical practice standards were relevant to carrying out each of the pipeline tasks (including working on a team in an ethical manner. We concluded, at the end of Section 2, that the entirety of these ethical practice standards is supportive of ethical practice with data (whether in statistics, data science, or computing). In Section 3 however, the vignettes tended to involve more of the practice standards in determining alternative decisions (KSA 4), and in their justifications (KSA5). One reason for this is that there are more stakeholders, and so more potential harms to prevent or offset, the further along the pipeline the work gets. It is important to recognize that not only is ethical practice important for every pipeline task – just to carry out each task ethically – but also, ethical decision-making at an earlier pipeline task could prevent unethical decisions or behaviors at later pipeline tasks.

The objective of Section 3 was to teach and give practice in the *full range of KSAs* that are required for ethical reasoning – rather than to make sure every conceivable situation was explored. The attentive reader will have noticed in Section 2 that working on teams ethically involved every principle of the ACM CE and the ASA GLs – irrespective of which of the other pipeline tasks are carried out while working on a team. Moreover, whenever KSAs 1-2 are needed – i.e., whenever engaging in any of the statistics and data science pipeline tasks – not only are the ACM and ASA ethical practice standards useful, but also workplace policies and local laws are, as well. You will not find that the ethical practice standards are in conflict with laws -with the specific exceptions of ACM 1.2, "To minimize the possibility of indirectly or unintentionally harming others, computing professionals should follow generally accepted best practices *unless there is a compelling ethical reason to do otherwise"* (emphasis added) and ASA A11, "Follows applicable policies, regulations, and laws relating to their professional work, *unless there is a compelling ethical justification to do otherwise."* (Emphasis added). Both practice standards recognize that there may be a point of practice where the practice standards conflict with either best practices (ACM) or policies/regulations/laws relating to professional work (ASA). Both practice standards articulate that a compelling ethical reason (ACM) or justification (ASA) are required to support prioritizing the standard over the rule. One of the sub-aims of Section 3 was to ensure that readers consider, and practice reasoning through, what they might do if faced with competing priorities of stakeholders or conflicts between practice standards and local rules or policies. The role-playing activities in Chapter 3.9 can be used to more fully explore how to identify, and then clearly articulate, a compelling ethical reason or justification to deviate from expected practice (ACM) or policy/law (ASA).

According to the National Academies of Sciences, Engineering, and Medicine (NAS, 2018), "data science spans a broad(er) array of activities that involve applying principles for data collection, storage, integration, analysis, inference, communication, and ethics". (p. 1) This report, by the Committee on Envisioning the Data Science Discipline,

> "… underscores the centrality of studying the many ethical considerations that arise as workers engage in data science. These considerations include deciding what data to collect, obtaining permissions to use data, crediting the sources of data properly, validating the data's accuracy, taking steps to minimize bias, safeguarding the privacy of individuals referenced in the data, and using the data correctly and without alteration. It is important that students learn to recognize ethical issues and to apply a high ethical standard." (p. 2).

One of the 2018 NAS action items (p.3) follows directly from the Committee's focus on the centrality of ethical professional practice in data science:

> *Recommendation 2.4: Ethics is a topic that, given the nature of data science, students should learn and practice throughout their education. Academic institutions should ensure that ethics is woven into the data science curriculum from the beginning and throughout.*

The hope for this book is that it supports your engagement in thoughtful, career spanning, engagement with the importance of ethical practice, and ethical reasoning, whenever you work with data. *Professionalism* is defined as "the skill, good judgment, and polite behavior that is expected from a person who is trained to do a job well." And "the conduct, aims, or qualities that characterize or mark a profession or a professional person." (Merriam-Webster).

[28]*Professional Identity* is defined as "…the sense of being a professional…the use of professional judgment and reasoning … critical self-evaluation and self-directed learning…" (Paterson et al. 2002:7)

> ▶   "Professional identity formation means becoming aware of … what values and interests shape decision-making." (Trede, 2012:163)
> ▶   "(S)tudents could learn more from their experiences if they were more explicitly guided to look out for certain aspects of professionalism and given further opportunities to discuss and critique their observations and experiences. "(Grace & Trede, 2011:12)

Data science is a discipline that has emerged at the intersection of computing and statistics – two disciplines with long standing guidance for ethical practice that feature professional integrity and responsibility. Both practice standards represent the perspectives of experienced professionals in their respective domains, but both organizations explicitly state that the guidelines apply to – should be utilized by – all who employ the domain in their work, irrespective of job title or training/professional preparation. Therefore, awareness of the ASA Ethical Guidelines for Statistical Practice – which are explicitly supportive of ethical practices in research that uses data whether "big" or small should be augmented with similar understanding of ethical practices relating to computing. The ACM Code directs "computing professionals" to focus attention on the positive and negative effects of their decisions – and emphasizes decisions that avoid or minimize harms. In the preamble, it is stated that computing professionals act responsibly when "consistently supporting the public good".

Like the ASA Guidelines, the ACM Code seeks to support ethical decision making, rather than describe what is wrong with aspects of practice. Given that both statistics and computing are essential foundations for modern statistical practice as well as data science, their ethical guidance should therefore be a starting point for the community as it contemplates what "ethical statistics and data science" looks like. The ASA Ethical Guidelines for Statistical Practice and ACM Code of Ethics are consensus-based guidelines.

---

[28] This material was adapted from: Tractenberg RE. (2020, August 15). Ten simple rules for integrating ethics into statistics and data science instruction. SocArXiv. https://doi.org/10.31235/osf.io/z9uej.

I hope this book will help readers move towards satisfying the expectations of "a person who is trained to do a job well", rather than satisfying a requirement of "awareness of ethical practice standards". As articulated by both the ASA and ACM, users of statistics and computing need to understand their responsibilities to practice with these bodies of disciplinary knowledge **ethically.**

These professional practice standards are maintained and updated over time as the profession changes. ASA Guidelines (2018) and the ACM Code of Ethics (2018) support professionalism, *viz* "…the skill, good judgment, and polite behavior that is expected from a person who is trained to do a job well." Failing to inform those who use/learn to use statistics and data science – at all levels - of their obligations to practice ethically does not make the responsibilities go away, but it does ensure that fewer practitioners are able to take responsibility for ethical practice.

Ethical reasoning is its own set of knowledge, skills, and abilities (**KSA**s (Santa Clara University (no date); Tractenberg & FitzGerald, 2012; Tractenberg et al. 2017)). These KSAs are learnable and improvable, and can be deployed to ensure ethical practice (when there is no/before there is an ethical problem about which a decision has to be made) as well as when a decision about what to do (ethically) is required. Thus, learning to reason ethically – rather than "learning the Ethical Guidelines and/or Code of Ethics" – will promote "…the skill, good judgment, and polite behavior that is expected from a person who is trained to do a job well" more generally, and more universally (see Rios et al. 2019).

As you are now well aware, there are seven tasks that all statistics and data science can/do follow/recognize, representing the statistics and data science pipeline (read even more about ethical practice along this pipeline in Tractenberg, 2022):

1. plan/design
2. collect/munge/wrangle data
3. analysis – literal for statistics & data science, "evaluation" for computing
4. interpret – always for statistics & data science, never for computing
5. document your work
6. report & communicate
7. work on a team

These tasks are essential in the practice of statistics and data science, and the ASA Ethical Guidelines pertain in <u>each</u> of these tasks; the ACM Code of Ethics pertains in almost all of them (but since computing and statistics are different, naturally their ethical practice standards are differently relevant in statistics and data science). Ethical aspects of practice by non-statisticians as well as statisticians are always relevant, and should be normalized. By this point, the reader should be able to make a case that knowing that there are Guidelines/a Code, and even what they contain, is not enough for ethical practice of statistics and data science or computing. Because different situations require different principles and elements of the Guidelines and Code, ethical reasoning is an important skill set that can be learned and improved. Ethical reasoning can be brought to bear on situations where the Guidelines or Code – or workplace policies/rules - may be useful. However, the ER KSAs can also be applied in situations not specifically related to statistics and data science or computing.

Since argument and justification with the ethical reasoning KSAs can be repeated with different material and different expectations as readers (you!) learn and improve their ethical reasoning abilities, you may find different nuances in your responses throughout your career. Clearly the ASA (Principle G) and ACM (Principle 3) focus on responsibilities that are specific to those in leadership roles will add different dimensions to even the cases explored in Section 3 over time, should a reader choose to revisit these vignettes.

Recall that there are six levels of complexity (from Bloom's 1 "B1" to Bloom's 6 "B6") in the original Taxonomy: **B1** Remember/Reiterate (performance is based on recognition of previously seen example); **B2** Understand/Summarize (performance summarizes info already in question (and/or answers)); **B3** Apply/Illustrate (performance extrapolates from seen examples to (really) new examples); **B4** Analyze/Predict (performance requires application/following of a rule; organize/sort using concrete criteria; some use of judgement in conjunction with the rules is needed); **B5** Create/ Synthesize (find patterns, innovate – to fill gaps in knowledge); and **B6** Evaluate/Compare/Judge (make judgments in the absence of "truth" or concrete criteria). The *minimum* level of performance to strive for, for ethical practice, is **B4**, as shown in Figure 3.10.1 (from Chapter 1.1).

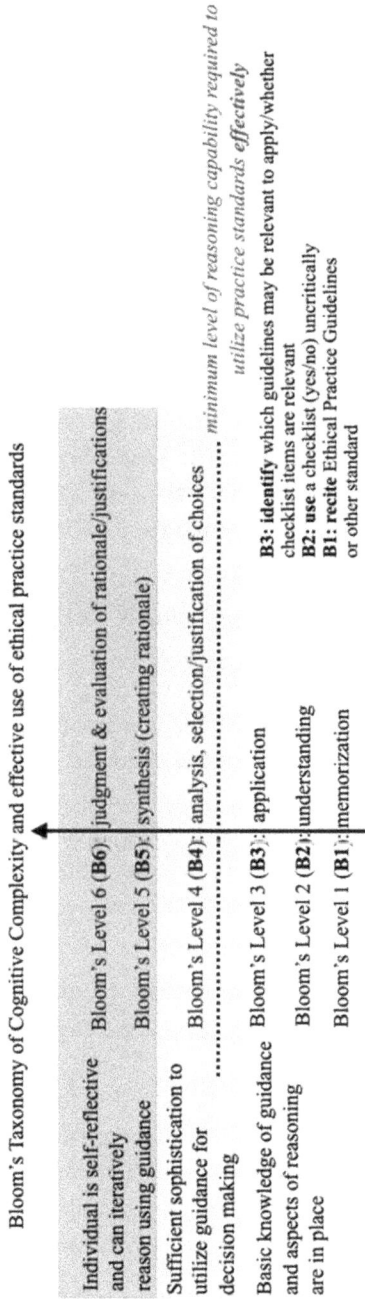

Figure 3.10.1: *Bloom's cognitive complexity level 4 is the minimum required to apply the judgment needed to utilize ethical guidance.*

The first time I used this figure, the label (and my argument) said, "**Memorization is not sufficient**". I wonder if you agree with this Figure after having read through the analyses and discussions in Sections 2 and 3.

"Learning ethics" is an oversimplified objective. The approach of this book is to *teach ethical reasoning*, specifically, how to make decisions throughout statistical and data science practice *ethically*. As you have seen, both the ASA and ACM intend for those to whom their ethical practice standards apply (i.e., any user of their tools, techniques, and technologies) to do so, and make decisions about doing so, in an ethical way. This requires the knowledge, skills, and abilities of ethical reasoning that the reader has now learned and practiced throughout the statistics and data science pipeline. At this point I hope it is clear that ethical reasoning is useful in interesting cases that were not covered in Section 3, for example, in dealing with issues like "privacy" or "algorithmic fairness". As we have seen in this book, ethical practice in statistics, data science, and computing involves much more than simply respecting privacy and fairness!

The important feature of ethical reasoning is that the emphasis is not on "the right answer", it is instead on how you, the practitioner, arrive at the answer/decision, and that the decision-making is evaluable and reproducible. Assumptions and background knowledge are recognized so that, if these change, then the decisions can be revisited -in equally evaluable ways- if needed. The focus in this book on ethical reasoning with the ASA and ACM professional practice standards has hopefully given you sufficient experience with ethical reasoning KSAs so that if other practice standards become relevant (e.g., in the workplace), the same approach can be used with the new material. I wonder if you agree with this.

Both the ASA, representing roughly 18,000 practitioners worldwide and the ACM, representing roughly 100,000 computing professionals worldwide, assert that their ethical practice guidance should pertain to members and non-members alike who utilize their methods and techniques. Thus, promoting the ethical use of statistical and computing practices through these ethical practice standards is justified (Martinson et al. 2005; Institute of Medicine, 2009; NASEM, 2017; Stark & Saltelli, 2018; Wang et al. 2018) and timely (National Academies of Sciences, Engineering, and Medicine, 2018).

The reader is encouraged to continue with self-directed learning, using the same KSAs on new problems – possibly with new or evolving practice standards or with workplace policies and rules and in particular, reflecting on the utility of each analysis to improve the chances of an *ethical* data-centered world.

To support self-directed growth in ethical reasoning, the reader is invited to consider, and discuss, how the ASA Guidelines and ACM Code of Ethics promote professionalism (in you) or the profession of statistics and data science (more generally).

- Explain whether/how the application of the Guidelines in any given case encourages ethical conduct in research (or practice, as appropriate) more generally.

- Do the ASA Guidelines and/or ACM Code of Ethics promote *professionalism*? How/how not?

- Discuss how the ASA Guidelines and ACM Code of Ethics promote professionalism (in you) or the profession of statistics and data science (more generally).

- Explain whether/how the application of the Guidelines in this case encourages ethical conduct in research (or practice, as appropriate)

- Do the ASA Guidelines or ACM Code of Ethics promote *professionalism* in these cases? How/how not?

- Why is "do nothing" a decision in any given case? Is it ever justifiable to "do nothing" or "do nothing differently"? How so/why not?

- Why should you do or say something (in a case like this) when no one else seems to think there is a problem?

- Why is "no one else seems to have a problem (with this)" NOT a good answer?

- What justification is there for a claim that "ethical practice is just common sense applied at work"? Is that true (in this case)?

- Why do people say that there is "no right answer" to problems/questions of ethical practice?

References

American Statistical Association (ASA) *ASA Ethical Guidelines for Statistical Practice-revised* (2022) downloaded from https://www.amstat.org/ASA /Your-Career/Ethical-Guidelines-for-Statistical-Practice.aspx on 1 January 2022.

Anderson RE, Johnson DG, Gotterbarn D, Perrolle J. (1993). Using the new ACM code of ethics in decision making. *Communications of the* ACM, 36(2): 98-107.

Arnold L & Stern DT. (2006). A framework for measuring professionalism. In DT Stern (Ed.), *Measuring Medical Professionalism*. New York, NY: Oxford University Press. Pp. 3-14.

Association for Computing Machinery (ACM). *Code of Ethics* (2018) downloaded from https://www.acm.org/about-acm/code-of-ethics on 12 October 2018.

Briggle, A., & Mitcham, C. (2012). *Ethics and science: An introduction*. Cambridge, UK: Cambridge University Press.

BS Bloom (Ed.), with Engelhart MD, Furst EJ, Hill WH & Krathwohl DR. (1956). *Taxonomy of educational objectives: Handbook I: Cognitive domain.* New York, NY: David McKay.

Campbell DT. III. "Degrees of Freedom" and the Case Study. *Comp Polit Stud.* 1975;8: 178–193. doi:10.1177/001041407500800204

Collins FS & Tabak LA. (2014). Policy: NIH Plans to enhance reproducibility. *Nature* 505 (30 Jan 2014): 612-613. Downloaded from http://www. nature.com/news/policy-nih-plans-to-enhance-reproducibility-1.14586 on 10 March 2016

Coventry LL, Finn J, Bremner AP. (2011). Sex differences in symptom presentation in acute myocardial infarction: a systematic review and meta-analysis. *Heart Lung* 40(6):477-91. doi: 10.1016/j.hrtlng.2011.05.001.

Ferrini-Mundy J. (2008). What core knowledge do doctoral students in mathematics education need to know? In RE Reys, JA Dossey (Eds). *U.S.*